Optoelectronics
Circuits Manual

Newnes Circuits Manual Series

CMOS Circuits Manual   R. M. Marston

Optoelectronics Circuits Manual

R. M. Marston

Heinemann:London

To Ashley, with love

Heinemann Professional Publishing Ltd
22 Bedford Square, London WC1B 3HH

LONDON   MELBOURNE   JOHANNESBURG   AUCKLAND

First published 1988

© R. M. Marston 1988

British Library Cataloguing in Publication Data
Marston, R. M.
  Optoelectronics circuits manual. –
  (Newnes circuits manual series).
  1. Electrooptical devices
  I. Title
  621.36   TA1750

ISBN 0 434 91211 5

Printed in England by
Redwood Burn Ltd, Trowbridge

Contents

Preface

Optoelectronics can be defined as the study of any devices that produce an electrically-induced optical (visible or invisible light) output, or an optically-induced electrical output, and of the electronic techniques and circuitry used for controlling such devices. It is one of the fastest-growing branches of modern electronics and encompasses a wide variety of devices, ranging from simple light bulbs and light-emitting diodes (LEDs) to complete infra-red light-beam alarm and remote control systems.

This book is intended to act as a useful single-volume guide to the optoelectronics device user, and is specifically aimed at the practical design engineer, technician, and experimenter, as well as the electronics student and amateur. It deals with the subject in an easy-to-read, down-to-earth, non-mathematical but very comprehensive manner. It starts off by explaining the basic principles and characteristics of the best known types of optoelectronic device, and then goes on to look at the practical applications of many of these devices in depth.

Individual chapters of the book are devoted to LED display circuits, to LED dot- and bar-graph circuits, to applications of seven-segment displays, light-sensitive devices, and optocouplers, and to a variety of brightness-control techniques. The final two chapters deal with infra-red light-beam alarms and with multichannel remote control systems.

Throughout the book, great emphasis is placed on practical user information and circuitry, and the book abounds with useful circuits and graphs; over two hundred and sixty diagrams are included. Most of the ICs and other devices used in the practical circuits are inexpensive and readily available types, with universally recognized type numbers.

1 Basic principles

Optoelectronics can be regarded as the study of any devices that produce an electrically-induced optical (visible or invisible light) output, or an optically-induced electrical output, and of the electronic techniques and circuitry used for controlling such devices. Optoelectronics is obviously a fairly large subject; in this chapter we present a brief survey of some of the devices, principles and techniques that it entails.

The best known types of light-generating optoelectric device are the ordinary tungsten filament lamp, the LED (light-emitting diode), the multisegment LED array, and the neon or gas-discharge lamp; other types of light-generator include the cathode ray tube and the LASER. An associated device is the LCD (liquid-crystal display), which does not in fact generate light but produces variations in the device's ability to reflect existing ambient light.

Light-sensitive devices include photodiodes and phototransistors (which have optosensitive conductivity), light-sensitive resistors (which have opto-sensitive resistivity), and so-called solar or photovoltaic cells (which are optosensitive voltage generators). Some specialist devices such as opto-isolators and optoreflectors combine both light-generating and light-sensitive units in a single package.

Optoelectronic devices have many practical applications. They can be used to generate a wide variety of stationary or moving visual displays. They can be used to give an automatic switching or alarm action in the presence or absence of a visible or invisible light source, or to give a similar action when a person or object moves within range of a generated light source. They can be used to give remote-control action via an infra-red light generator and a remotely placed detector. In some applications, fibre-optic cables can be used to form a low-loss closed-circuit connecting link between a code-modulated light generator and a matching remotely placed light-sensitive device, thus forming an interference-free data link. We will take a brief look at all of these devices and techniques in the next few pages.

Filament lamps

The simple filament lamp or light bulb is the best known type of light generator. It is widely used in the home, the car, and in industry, can be powered from either AC or DC voltage sources, and uses the standard circuit symbol of *Figure 1.1(a)*. It usually comprises a coil of tungsten wire (the filament) suspended within a vacuum-filled glass envelope and connected to the outside world via a pair of metal terminals; the filament runs white hot when connected to a suitable external voltage, thus generating a bright white light.

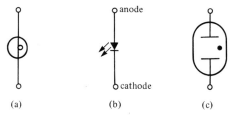

(a)   (b)   (c)

Figure 1.1   *Circuit symbols of (a) filament lamp, (b) LED, and (c) neon lamp*

The filament lamp has two notable characteristics. One of these is that its resistance varies with filament temperature, and *Figure 1.2* shows the typical variation that occurs in a 12 V 12 W lamp. Thus, the resistance is 12 Ω when the filament is operating at its normal 'white' heat, but is only 3 Ω when the

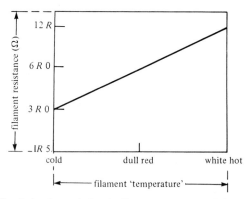

Figure 1.2   *Graph showing variations in filament resistance with filament temperature for 12 V 12 W lamp.*

filament is cold. This 4:1 resistance variation is typical of all filament lamps, and causes them to have switch-on 'inrush' current values about four times greater than the normal 'running' values.

The other notable feature of the filament lamp is that it has a fairly long thermal time constant, so that power has to be applied to (or removed from) the filament for a significant time (tens or hundreds of milliseconds) before it has any appreciable effect on light output. This characteristic enables the device to be powered from either AC or DC voltage sources, and enables the lamp brightness to be varied by using highly efficient switched-mode 'pulsing' techniques.

Light-emitting diodes

Another well known type of light-generating device is the light-emitting diode (LED). *Figure 1.1(b)* shows the standard circuit symbol of this solid-state device, which has electrical characteristics similar to those of a normal diode (i.e., it passes current in one direction and blocks it in the other), but emits light when biased in the forward direction. Standard types of LED emit a red coloured light, but other types are available that emit orange, yellow, green, or infra-red types of light.

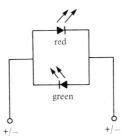

Figure 1.3   *Bi-colour LED shows red or green*

LEDs require typical forward operating voltages of about 2 V and forward currents of 10 to 20 mA. They are widely available in single-LED packages, but are also available in multi-LED styles. 2-LED packages housing a pair of red and green coloured LEDs are, for example, available in either bi-colour or tri-colour forms, as shown in *Figures 1.3* and *1.4*. In the bi-colour device only one LED can be illuminated at a time, so the device emits either a red or green colour, but in the tri-colour device both LEDs can be illuminated at the same time, generating a yellow colour in addition to red and green. Multi-LED packages are also available in bargraph form (*Figure 1.5* shows the circuit of a ten-LED bar-graph device), in 5 by 7 dot matrix form, and in seven-segment display form (described later).

4   Basic principles

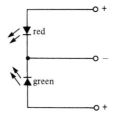

Figure 1.4   *Tri-colour LED shows red, green or yellow (with both LEDs on)*

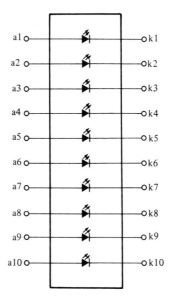

Figure 1.5   *Circuit of ten-LED bargraph display*

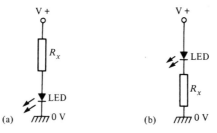

Figure 1.6   *LED currents can be limited via a series resistor connected to either the anode (a) or the cathode (b)*

In use, the operating current of a LED must be limited to a safe value; this can be achieved via a series resistor connected to either the anode or the cathode, as shown in *Figure 1.6*. Note that LEDs have very brief optoelectric response times, and can easily be used to transmit coded remote control light signals, etc.

Neon lamps

Neon gas discharge lamps can also be used as light-generating devices. They comprise a neon gas and a pair of electrodes housed in a glass envelope. When a suitably high striking voltage is applied to the electrodes the gas becomes conductive, producing a red glow on the electrodes; if the voltage is further increased the glow spreads through the neon gas. In use, a resistor is wired in series with the neon lamp, so that the neon voltage self-limits to slightly above the striking value. *Figure 1.1(c)* shows the circuit symbol of the neon lamp, which can be powered from either an AC or DC voltage.

Fluorescent displays

Another type of light-generating device is the fluorescent or phosphorescent display, which is shown in basic form in *Figure 1.7*. Here, an incandescent filament (typically using a 2 V supply) acts as a source of free electrons, which

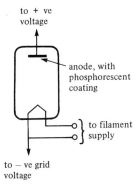

to + ve
voltage

anode, with
phosphorescent
coating

to filament
supply

to − ve grid
voltage

Figure 1.7   *Basic construction of a fluorescent or phosphorescent display device*

can be accelerated into a phosphor-coated anode via a suitable grid-to-anode voltage (typically about 24 V), thus generating a visible green or blue fluorescent glow. This type of device is available in seven-segment display form (described shortly).

Liquid crystal displays

The four basic types of optoelectric display device that we have just looked at inevitably consume substantial electric power, since they actually generate light. Liquid crystal displays (LCDs), on the other hand, are used to reflect existing ambient light, and can thus operate with negligible power consumption. *Figure 1.8* shows the basic structure of an LCD device designed to display either a blank or the digit 1.

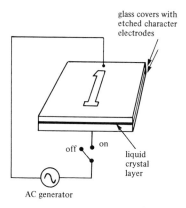

glass covers with
etched character
electrodes

off
on
liquid
crystal
layer

AC generator

Figure 1.8   *Basic structure of a LCD device designed to display either a 1 or a blank*

Here, the display device consists of a very thin layer of liquid crystal, sandwiched between two glass covers which have the transparent character 1 etched on to them in the form of an externally-available pair of electrodes. Normally, the liquid crystal molecules are randomly aligned, and the complete unit appears as a simple (blank) block of transparent glass. When, on the other hand, an AC voltage (usually 40–100 Hz) is applied across the 1-shaped electrodes, the molecules within the intervening layer of liquid crystal

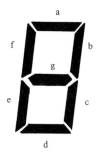

a
f
b
g
e
c
d

Figure 1.9   *Standard form and notations of a seven-segment display*

become agitated, taking up a mirror-like optical density that vividly reflects the character 1 from any existing ambient light source. The device reverts to its blank state when the AC excitation voltage is removed.

In reality, the etched character shape (or shapes) of an LCD device can take any desired form, and in practice they are most widely used in the form of seven-segment displays. Note that LCDs are voltage-operated devices, and consume near-zero quiescent power.

Seven-segment displays

A very common requirement in modern electronics is that of displaying alphanumeric characters: digital watches, pocket calculators, and digital instruments are all examples of devices that use such displays. The best known display of this type is the so-called seven-segment display; it comprises seven independently-accessible photoelectric segments (e.g., LED, LCD, gas-discharge, fluorescent, or filament-type segments) arranged in the form shown in *Figure 1.9*. The segments are conventionally notated from a to g in the manner shown in *Figure 1.9*, and it is possible to make them display any

segments						(✓ = ON)	display	segments						(✓ = N)	display
a	b	c	d	e	f	g		a	b	c	d	e	f	g	
✓	✓	✓	✓	✓	✓		*0*	✓	✓	✓	✓	✓	✓	✓	*8*
	✓	✓					*1*	✓	✓	✓			✓	✓	*9*
✓	✓		✓	✓		✓	*2*	✓	✓	✓		✓	✓	✓	*A*
✓	✓	✓	✓			✓	*3*			✓	✓	✓	✓	✓	*b*
	✓	✓			✓	✓	*4*	✓			✓	✓	✓		*C*
✓		✓	✓		✓	✓	*5*		✓	✓	✓	✓		✓	*d*
✓		✓	✓	✓	✓	✓	*6*	✓			✓	✓	✓	✓	*E*
✓	✓	✓					*7*	✓				✓	✓	✓	*F*

Figure 1.10 *Truth table for a seven-segment display*

numeral from 0 to 9 or any alphabetic character from A to F (in a mixture of upper and lower case letters) by activating these segments in various combinations, as shown in the truth table of *Figure 1.10*. A wide variety of digital IC types are available for providing suitably decoded driving signals for seven-segment displays.

Photodetectors

Photodetectors are devices that provide a change in electrical characteristics in the presence of a change in light input. The best known of these devices are the LDR (light-dependent resistor), the photodiode, and the phototransistor, and *Figure 1.11* shows the circuit symbols of these three devices.

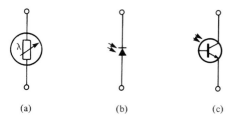

(a)   (b)   (c)

Figure 1.11   *Circuit symbols of (a) LDR, (b) photodiode, and (c) phototransistor*

The LDR is also known as a cadmium-sulphide (CdS) photocell; it is a passive device that simply changes its electrical resistance in the presence of external light. *Figure 1.12* shows the typical photoresistive graph that applies to an LDR with a face diameter of about 10 mm.

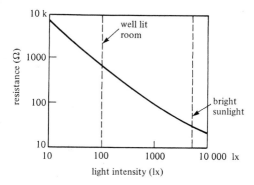

Figure 1.12   *Typical photoresistive graph of LDR with 10 mm diameter face*

A photodiode is a normal silicon diode that is either mounted in a translucent case or has its semiconductor junction mounted beneath an optical lens. It is a simple fact of life that if any silicon diode junction is reverse biased in the circuit of *Figure 1.13*, its reverse-current value will depend on the amount of illumination on the junction face, being near-zero under dark conditions and tens or hundreds of nA under bright conditions. Similarly, a phototransistor is a normal silicon transistor with a photovisible junction; it

has a far greater sensitivity than the photodiode, and can be made to act as a
sensitive light-to-voltage converter by wiring it in either of the configurations
shown in *Figure 1.14*.

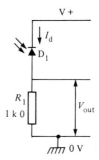

Figure 1.13 *Basic photodiode circuit*

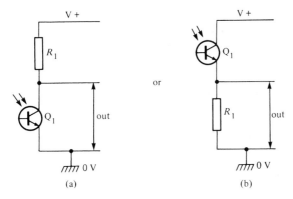

(a)   (b)

Figure 1.14 *Alternative phototransistor configurations*

Figure 1.15 *Symbol of a single solar cell*

Solar cells

So-called solar cells are actually photovoltaic units that convert light directly
into electrical energy, *Figure 1.15* shows the symbol used to represent a single
solar cell.

10 Basic principles

An individual solar cell generates an open circuit voltage of about 500 mV (depending on light intensity) when active. Individual cells can be connected in series to increase the available terminal voltage, or in parallel to increase available output current; banks of cells manufactured ready-wired in either of these ways are known as solar panels. *Figure 1.16* shows how a bank of 16 to 18 cells can be used to autocharge a 6 V ni-cad battery via a germanium diode.

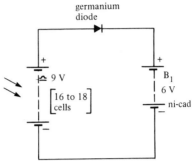

Figure 1.16   *Solar panel used to charge a 6 V ni-cad*

The available output current of a solar cell depends on the light intensity, on cell efficiency (typically only a few per cent), and on the size of the active area of the cell face. Note that available sea-level light energy is typically in the range 0.5 to 2 kW/m² on a bright sunny day, so there is plenty of 'free' energy waiting to be converted!

Optocouplers

An optocoupler is a device containing an infra-red LED and a matching phototransistor, mounted close together (optically coupled) within a light-excluding package, as shown in the basic circuit of *Figure 1.17*. Here, SW_1 is

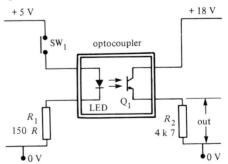

Figure 1.17   *Basic optocoupler circuit*

normally open, so zero current flows through the LED; Q_1 is thus in darkness and also passes zero current, so zero output voltage appears across R_2. When SW_1 is closed, however, current flows through the LED via R_1, thus illuminating Q_1 and causing it to generate an R_2 output voltage. The R_2 output voltage can thus be controlled via the R_1 input current, even though R_1 and R_2 are fully isolated electrically. In practice, the device can be used to optocouple either digital or analogue signals, and can provide hundreds or thousands of volts of isolation between the input and output circuits.

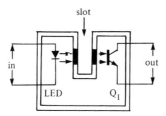

Figure 1.18 *Slotted optocoupler*

Figures 1.18 and *1.19* show two useful variants of the basic optocoupler. The first of these is the slotted optocoupler, which has a slot moulded into the package between the LED and Q_1, as shown in *Figure 1.18*; the slot houses transparent windows, so that the LED light can normally freely reach the face of Q_1, but can be interrupted or blocked via an opaque object placed within the slot. The slotted optocoupler can thus be used as an object detector.

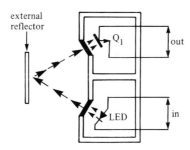

Figure 1.19 *Reflective optocoupler*

The second device is the reflective optocoupler shown in *Figure 1.19*. Here, the photoactive faces of the LED and Q_1 both point outwards (via transparent windows) towards an imaginary point that is roughly 5 mm beyond each window, so that the LED light can only reach Q_1 face via a reflective surface that is placed at or near this point. This device can thus also be used as an external-object detector.

Light-beam systems

One of the most important applications of the infra-red LED/phototransistor combination is in the making of light-beam systems, which can include light-beam alarms, infra-red remote-control systems, and (with the aid of fibre-optic cables) infra-red data links. The basic principles of these systems are illustrated in *Figures 1.20* to *1.24*.

Figure 1.20 shows the basic operating principle of the simple light-beam alarm. Here, the transmitter feeds a coded signal (usually a fixed-tone square wave) into an infra-red LED, which has its output focused into a 'beam' that is aimed at a matching infra-red phototransistor mounted on the remotely placed receiver. The circuit action is such that (when the light-beam is operating) the receiver output is normally off, but automatically activated an external alarm, counter, or relay if the beam is interrupted by a person, animal or object. This type of system can have an effective detection range of up to 30 metres.

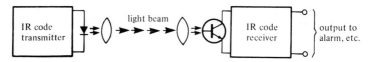

Figure 1.20   *Simple light-beam alarm*

The above system works on the pin-point 'line of-sight' principle, and can be activated by any 'bigger-than-a-pin' object that enters the line-of-sight between the transmitter and receiver lenses. Thus, this system can easily be false-triggered by a fly or moth (etc.) entering the beam or landing on one of the lenses: The dual-light-beam system of *Figure 1.21* does not suffer from this defect.

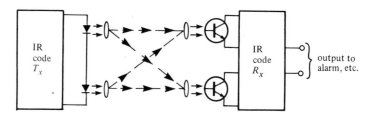

Figure 1.21   *Dual-light-beam alarm*

The *Figure 1.21* system is basically similar to that already described, but transmits the infra-red beam via two series-connected LEDs that are normally spaced about 75 mm apart, and receives the beam via two parallel-connected phototransistors that are also spaced 75 mm apart. Thus, each phototransistor can detect the beam from either LED, and the receiver will thus not activate if one or other of the beams is broken; the receiver will only activate if *both* beams are broken simultaneously, and this will normally only occur if a large (greater than 75 mm) object is placed within the composite beam. This system is thus virtually immune to false triggering by moths, etc.

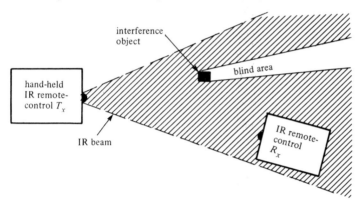

Figure 1.22   *Basic IR remote-control system*

Figure 1.22 illustrates the operating principle of an infra-red remote-control system. Here, the hand-held unit transmits a broad beam of coded infra-red light, and can remote-control a receiver that is placed anywhere within the active area of this beam. Note that the transmitter and receiver do not need to be pointed directly at each other to effect operation, but *must* be in line-of-sight contact; also note that an object placed within the beam can create a blind area in which line-of-sight contact cannot exist.

Code waveforms

Infra-red LEDs and phototransistors are very fast acting devices; consequently, the effective range of an infra-red beam system is determined by the peak current fed into the transmitting LED, rather than by the mean transmitting current. These are important points to note when designing beam code waveforms, as illustrated in *Figure 1.23*.

The simplest type of code waveform is the fixed tone square wave signal, as shown in *Figure 1.23(a)*. Here, the mean transmitting current is half the peak value, so this system is not very efficient. Better efficiency is shown by the pulse system of *Figure 1.23(b)*, which transmits a 100 μS pulse once every 1100 μS, and thus has a mean current consumption that is only 1/10th of the peak value. Finally, the most efficient system of all is the tone-burst system

14 Basic principles

(*Figure 1.23(c)*) which, in the example shown, consumes a mean current that is only 1/200th of the peak value. Note that this latter system transmits a 1 mS sample burst of 10 kHz tone once every 100 mS (e.g., ten times per second), and is used mainly in light-beam intruder detecting systems.

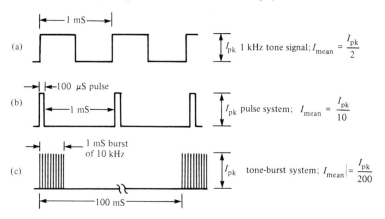

Figure 1.23 *Basic types of IR-beam code waveform*

Most multichannel remote-control systems use a fairly complex beam coding system, in which a number of bits of data are sequentially transmitted within a repeating frame, which also repeats many times per second. Thus, in each frame, the first bit may contain data for channel 1, the second for channel 2, the third for channel 3, and so on. In practice, 32-channel infra-red remote-control systems are readily available.

Fibre-optics

Fibre optic cables can, in very simple terms, be regarded as flexible light pipes that can efficiently carry modulated or unmodulated light signals from one point to another (even if the journey involves bends and loops) with little signal loss and complete immunity from electromagnetic interference. The simplest application of such cables is in distributing the visible light of a single source to many different pin point locations, as in, for example, a vehicle's instrument panel; in this instance the cables need no special treatment, and can simply be cut to length with a sharp knife. In more complex applications, such as the coded data link of *Figure 1.24*, the cable needs to be united with the light source and the distribution point via special connectors, to cut down signal losses.

Two distinct types of fibre-optic cable are in common use. One of these is an inexpensive type made from polymer cable; it is easily cut, ideally suited for use with visible red light, and is best suited to short-distance (up to 10 m) applications; it gives a maximum attenuation of 200 dB/km of cable. The other type of cable uses a glass fibre construction; it is expensive and difficult to cut, but can efficiently handle infra-red signals and exhibits a typical transmission loss of only 5 dB/km.

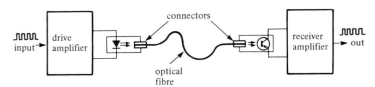

Figure 1.24 *Basic fibre optic link*

The cathode ray tube

No introductory survey of optoelectronics would be complete without at least a brief mention of the cathode ray tube, which is widely used as a TV screen, as a computer monitor screen, and as an oscilloscope display unit. *Figure 1.25* illustrates the basic structure of the oscilloscope version of such a unit.

In *Figure 1.25*, the cathode is heated by a filament and acts as a source of free electrons, which sharply accelerate along the tube towards the positively-charged anode and are steered towards its centre by a grid voltage. The anode, however, has a pin-point hole in its centre, so a narrow beam (the cathode ray) of fast-moving electrons speeds on to the end of the tube (the screen), which is internally coated with phosphor powder and thus produces a pin-point glow of light as it is struck by the beam.

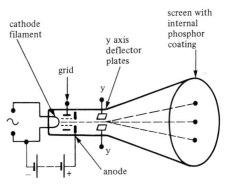

Figure 1.25 *Basic structure of a cathode ray tube*

In *Figure 1.25* it can be seen that the cathode ray electron beam can actually be moved up or down (on the Y axis) by applying a suitable voltage to an internal pair of deflection plates. In practice, the tube contains two pairs of such plates, as shown in the frontal view of *Figure 1.26*; the X plates enable the beam to be deflected to the left or right, and the Y plates enable it to be deflected up or down. Thus, the bright spot can be moved to any part of the screen via the X and Y plates, which (by using rapid movements) can thus be used to draw an infinite variety of patterns, shapes or numerals, etc., on the screen.

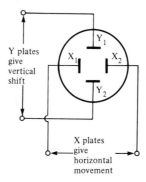

Figure 1.26   *Layout of X and Y deflection plates*

Note here that the beam deflection of the oscilloscope-type cathode ray tube described above is caused electrostatically, by the voltages applied to the X or Y plates. A similar deflection can be obtained electromagnetically, by feeding suitable currents to sets of coils mounted on the outside of the tube; this latter technique is in fact used with most TV and monitor tubes. In practice, most cathode ray tubes can also be controlled on the Z or beam-intensity (brightness) axis by feeding suitable control voltage signals to the grid terminal.

Vacuum photocells

Finally, to complete this introductory survey of optoelectronic devices, brief mention must be made of the vacuum photocell and the photomultiplier. The vacuum photocell (*Figure 1.27(a)*) consists of an anode and a photocathode mounted in a vacuum-filled transparent glass tube. In use (*Figure 1.27(b)*), the anode is heavily biased positively (to 100–500 V). Each time the cathode is hit by a light particle (a photon) it releases an electron, which is collected by the anode, thus resulting in a light-dependent current flow. This current has a typical 'illuminated' value of only a few microamps.

Photomultipliers work in a similar way to the vacuum photocell, except that the electrons reach the anode via a series-connected set of positively biased surfaces called dynodes (see *Figure 1.28*), which each release several electrons each time they are struck by one electron. Thus, if the photo-multiplier uses five dynodes which each give a multiplication factor of four, the dynodes will give progressive outputs of 4, 16, 64, 256 and 1024 electrons each time that the photocathode releases one electron, thus giving an overall sensitivity about one thousand times greater than the simple vacuum diode.

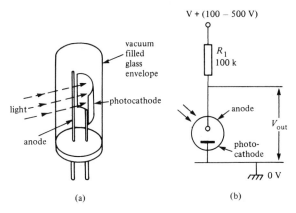

Figure 1.27   *Construction (a) and typical circuit (b) of vacuum photocell*

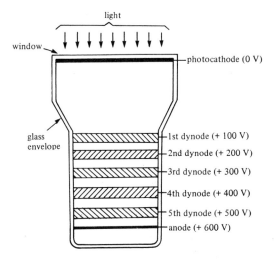

Figure 1.28   *Basic construction of a five-stage photomultiplier*

2   LED display circuits

The best known of all optoelectronic devices is the so-called LED (light-emitting diode), which emits a fairly narrow bandwidth of visible (usually red, orange, yellow or green) or invisible (infra-red) light when its internal diode junction is stimulated by a forward electric current/voltage (power). LEDs have typical power-to-light energy conversion efficiencies some ten to fifty times greater than a simple tungsten lamp and have very fast response times (about 0.1 μS, compared with tens or hundreds of milliseconds for a tungsten lamp), and are thus widely used as visual indicators and as moving-light displays; a variety of such circuits are shown in this chapter.

LED basics

Figure 2.1 shows the standard symbol that is used to represent the LED, which is a genuine diode and has a significant voltage (roughly 2 V) developed across it when it is passing a forward current; *Figure 2.2* shows the typical forward voltages of different coloured standard LEDs at forward currents of 20 mA. If a LED is reversed biased it will avalanche or zener at a fairly low voltage value, as shown in *Figure 2.3*; most practical LEDs have maximum reverse voltage ratings in the range 3 to 5 V.

A (anode)

K (cathode)

Figure 2.1   *LED symbol*

18

colour	red	orange	yellow	green
V_f (typical)	1 V 8	2 V 0	2 V 1	2 V 2

Figure 2.2   *Typical forward voltages of standard LEDs at $I_f = 20$ mA*

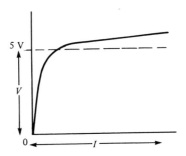

Figure 2.3   *A reverse-biased LED acts like a zener diode*

In use, a LED must be wired in series with a current-limiting device such as a resistor: *Figure 2.4* shows how to work out the R value needed to give a particular current from a particular supply voltage. In practice, R can be connected to either the anode or the cathode of the LED. The LED brightness is proportional to the LED current; most LEDs will operate safely up to absolute maximum currents of 30 to 40 mA. A LED can be used as an indicator in an AC circuit by wiring it in inverse parallel with a normal diode, as shown in *Figure 2.5*, to prevent the LED being reverse biased; for a given brightness, the R value should be halved relative to that of the DC circuit.

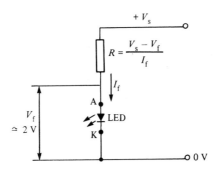

Figure 2.4   *Method of finding the R value for a given V_s and I_f*

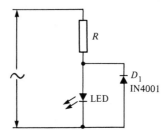

Figure 2.5   *Using a LED as an indicator in an AC circuit*

Practical usage notes

The first practical problem that will be met when using a LED is that of identifying its polarity. Most LEDs have their cathodes identified by a notch or flat on the package, or by a short lead, as indicated in the outline diagram of *Figure 2.6*. This practice is not universal, however, so the only sure way to identify a LED is to test it in the basic circuit of *Figure 2.4*; try the LED both ways round: when it glows, the cathode is the most negative of the two terminals. Note that it is always good practice to test a LED before soldering it into circuit.

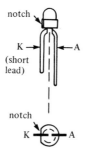

Figure 2.6   *Typical outline and method of recognizing the polarity of a LED*

Special mounting kits, comprising a plastic clip and ring, are available for fixing LEDs into PC boards and front panels, etc. *Figure 2.7* illustrates the functioning of such a kit.

Most LEDs come in the form of a single-LED package of the type shown in *Figure 2.6*. Multi-LED packages are also available, however. The best known of these are the seven-segment displays, comprising seven (or eight) LEDs packaged in a form suitable for displaying alphanumeric characters. So-called bar-graph displays, comprising ten to thirty linearly-mounted LEDs in a single package, are also available.

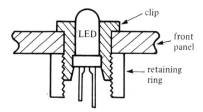

Figure 2.7   *Clip and ring kit used to secure a LED to a front panel, etc.*

Most LEDs provide only a single output colour. A few specialist devices do, however, provide multicolour outputs. These are actually two-LED devices, and *Figure 2.8* shows one such device that comprises a pair of LEDs connected in inverse parallel, so that the colour green is emitted when the device is biased in one direction, and red (or yellow) is emitted when it is biased in the reverse direction. This device is useful for giving polarity indication or null detection.

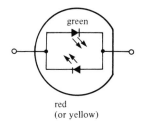

Figure 2.8   *Bi-colour LED actually houses two LEDs connected in inverse parallel*

Another type of multicolour LED is shown in *Figure 2.9*. This comprises a green and a red LED mounted in a three-pin common-cathode package. This device can generate green or red colours by turning on only one LED at a time, or can generate orange and yellow ones by turning on the two LEDs in the ratios shown in the table.

A very important practical point concerns the use of second grade or out-of-spec devices advertised as bargain packs. These devices often have forward volt drops in the range 3 to 10 V, and may thus be quite useless in many practical applications. *Always* test these devices before use.

22 LED display circuits

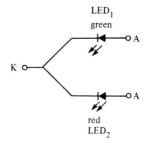

output colour	red	orange	yellow	green
LED₁ CURRENT	0	5 mA	10 mA	15 mA
LED₂ CURRENT	5 mA	3 mA	2 mA	0

Figure 2.9 *Multicolour LED, giving four colours from two junctions*

Multi-LED circuits

If you ever need to drive several LEDs from a single power source this can be done by wiring all LEDs in series, as shown in *Figure 2.10*. Note that the supply voltage used here must be significantly greater than the sum of the individual LED forward voltages. This circuit thus draws minimal total current, but is limited in the number of LEDs that it can drive. A number of

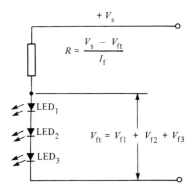

Figure 2.10 *LEDs wired in series and driven via a single current-limiting resistor*

these circuits can, however, be wired in parallel, so that almost any number of LEDs can be driven from a single source, as shown in the six-LED circuit of *Figure 2.11*.

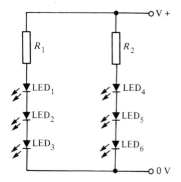

Figure 2.11   *Any number of* Figure 2.10 *circuits can be wired in parallel, to drive any number of LEDs*

An alternative way of simultaneously powering several LEDs is to simply wire a number of the *Figure 2.4* circuits in parallel, as shown in *Figure 2.12*. Note, however, that this circuit is very wasteful of current (which is equal to the sum of the individual LED currents).

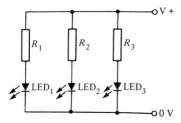

Figure 2.12   *This circuit can drive a large number of LEDs, but at the expense of high current*

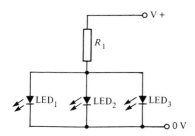

Figure 2.13   *This LED-driving circuit will not work. One LED will hog all the current*

Figure 2.13 shows a 'what *not* to do' circuit. This design will not work correctly because inevitable differences in the forward voltage characteristics of the LEDs will usually cause one LED to 'hog' most of the available current, leaving little or none for the remaining LEDs.

LED-control circuits

The three most widely used types of visible-output LED-control circuits are (ignoring those used for alphanumeric LED control) those used for LED flashing, for LED sequencing, and for LED dot or bar analogue-value indication.

LED flasher circuits are designed to turn a LED repeatedly on and off, to give an eye-catching display action; they may control a single-LED or may be designed to control two LEDs in such a way that one turns on as the other turns off and vice versa. A special LED-flasher IC is available (the LM3909) which can be used to flash a LED from a very low voltage (1 V 5) battery supply, and to do so at a very low mean current level. A dozen practical LED-flasher circuits are shown later in this chapter.

LED sequencer circuits are designed to drive a chain of LEDs in such a way that each LED in the chain is switched on and off in a time-controlled sequence, so that a ripple of light seems to run along the chain. Eleven such circuits are shown here.

LED analogue-value indicator circuits are designed to drive a chain of linearly-spaced LEDs in such a way that the length of chain that is illuminated is proportional to the analogue value of a voltage applied to the input of the driver circuit, e.g., so that the circuit acts like an analogue voltmeter. A range of practical analogue-value indicator circuits are shown in Chapter 3.

Simple LED-flasher circuits

One of the simplest types of LED display circuit is the LED flasher, in which a single LED repeatedly switches alternately on and off, usually at a rate of one or two flashes per second. A two-LED flasher is a simple modification of this circuit, but is arranged so that one LED switches on when the other switches off, or vice versa.

Figure 2.14 shows the practical circuit of a two-transistor two-LED flasher, which can be converted to single-LED operation by simply replacing the unwanted LED with a short circuit. Here, Q_1 and Q_2 are wired as a one cycle-per-second astable multivibrator, with switching rates controlled via C_1–R_3 and C_2–R_4.

Figure 2.15 shows an IC version of the two-LED flasher. This design is based on the faithful old 555 timer chip or its more modern CMOS counterpart, the 7555. The IC is wired in the astable mode, with its time constant determined by C_1 and R_4. The action here is such that output pin 3 of

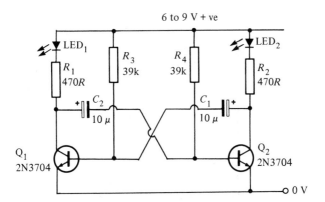

Figure 2.14 *Transistor two-LED flasher circuit operates at about 1 flash-per-second*

the IC alternately switches between the ground and the positive supply voltage levels, alternately shorting out or disabling one or other of the two LEDs. The circuit can be converted to single-LED operation by omitting the unwanted LED and its associated current-limiting resistor.

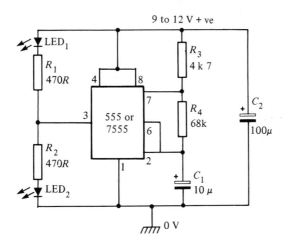

Figure 2.15 *IC two-LED flasher circuit operates at about 1 flash per second*

Figure 2.16 shows a useful modification of the above circuit, in which the flashing rate is made variable via RV_1, and two pairs of series-connected LEDs are connected in the form of a cross so that the visual display alternately switches between a horizontal bar (LED_1 and LED_2 ON) and a vertical bar (LED_3 and LED_4 ON), thus forming a visually interesting display. The flash rate is variable between 15 and 2000 per minute.

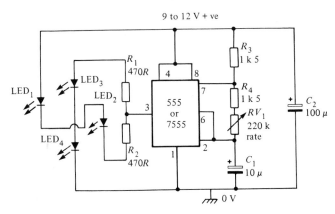

Figure 2.16   *Four-LED double-bar flasher. Rate is variable from 15 to 2000 flashes per minute*

The LM3909 flasher IC

A seemingly trivial task that sometimes faces the design engineer is that of providing an illuminated (glowing or flashing) indication of the ON state of a piece of electronic equipment or the location of a passive device (such as a fire extinguisher, torch, emergency switch, etc.) in a darkened room.

These tasks are obviously easily solved if mains power is available, but can present serious problems when battery powered equipment is concerned. LED indicators typically draw about 12 mA when illuminated and can thus place a fairly heavy strain on small supply batteries. LEDs, in any case, drop two or more volts under the ON condition and can thus not readily be powered from battery voltages below 3 V or so.

National Semiconductors provided an ingenious solution to this problem some years ago when they introduced the eight-pin LM3909 LED flasher/oscillator IC. This device acts basically as a low-duty-cycle (brief ON period, long OFF period) oscillator that provides a voltage-doubled high-current pulse to an external LED. Because of the voltage-doubling facility, this IC can flash a LED even when powered from cell voltages down to 1.1 V.

Because of the low duty cycle facility, the device can provide high pulse currents (up to 100 mA) while still drawing very low *mean* currents (typically 0.3 to 1.5 mA) and can thus provide months (or even years) of continuous flasher operation from a single 1.5 V cell.

The internal circuit of the LM3909, together with typical external connections for 1.5 V flasher operation, is shown in *Figure 2.17*. In this application the LED receives current (via the 270 μ capacitor and internal 12R resistor and Q_3) for only about 1% of the time. For the remaining part of each operating cycle all transistors except Q_4 are off. The internal 20 k resistor

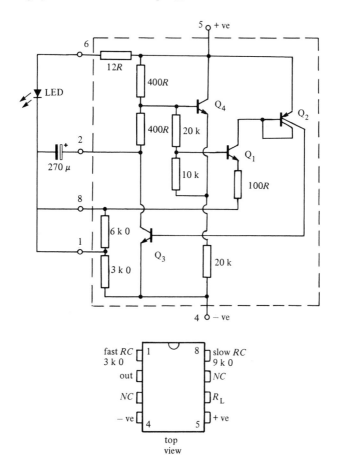

Figure 2.17 *Internal circuit and practical connections of the LM3909 low-voltage LED flasher. The IC outline is shown*

(from Q_4 emitter to supply common) draws only 50 μA or so. The 270 μ capacitor is charged via the two 400R resistors and (in this particular application) through the 3k0 resistor connected between pins 1 and 4 of the circuit.

Transistors Q_1 through Q_8 remain off until the 270 μ capacitor becomes charged to about 1 V. This voltage is determined by the junction drop of Q_4, its base-emitter voltage divider, and the Q_1 junction drop. When the pin 1 voltage becomes 1V0 more negative than that at pin 5 (the supply voltage pin) Q_1 begins to conduct and then turns Q_2 and Q_3 on. The IC then supplies a high-current pulse to the LED. The current amplification factor of Q_2–Q_3 is roughly 500; Q_3 can handle over 100 mA of collector current, and rapidly pulls pin 2 close to supply common (pin 4). Since the 270 μ capacitor is charged at this time, it forces the pin 1 terminal *below* the supply common value; consequently, the LED volt drop is greater than the supply voltage value! The internal 12R resistor (between pins 5 and 6) limits the LED current to a safe value.

Thus, in this particular application the 270 μ capacitor alternately charges via the 3k0 timing resistor and discharges via the LED and the internal 12R resistor. In some other applications the short between pins 1 and 8 can be removed, enabling the capacitor to charge through a total of 9k0, with a consequent increase in the duty cycle and reduction in mean current consumption. If voltage boosting is not needed (with or without current limit), loads can be wired directly between pins 2 and 6 or pins 2 and 5 of the IC. The LM3909 is thus a fairly versatile device, and the following section shows ten different ways of using it in flasher applications.

Practical LM3909 circuits

Figure 2.18 shows the *Figure 2.17* 1V5 flasher circuit redrawn in a practical configuration. The circuit gives a brief flash once every second or so and typically draws an average current of only 0.63 mA. As shown in the table, this circuit will give 3 to 30 months of continuous operation from a battery, depending on the size and type of cell that is used.

An even longer life can be obtained from the minimum power flasher circuit of *Figure 2.19*. This circuit is similar to the above, except that the short is removed from pins 1 and 8, causing the capacitor to charge via the 9k0 of internal IC resistance and so operate with an increased duty cycle and reduced mean current consumption; the circuit has a typical current drain of 0.32 mA.

The *Figure 2.18* and *2.19* circuits are of particular value as indicator or locator beacons for use in fire extinguishers, emergency lanterns, torches, and emergency switches, etc. The operating frequencies of these circuits are fairly heavily dependent on supply voltage, as is implied by the circuit of *Figure*

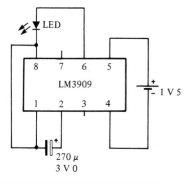

Figure 2.18   *Practical 1V5 LED flasher, with details of estimated battery life*

cell size	estimated battery life under continuous operation	
	standard cell	alkaline cell
A A	3 MONTHS	6 MONTHS
C	7 MONTHS	15 MONTHS
D	15 MONTHS	30 MONTHS

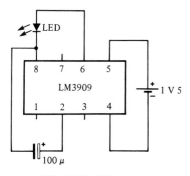

Figure 2.19   *Minimum power 1V5, 1.1 Hz LED flasher. Mean drain current is 0.32 mA*

2.20. This circuit is similar to that of *Figure 2.19*, except that it is designed for 3 V operation, in which case the timing capacitor value has to be increased by a factor of 2.7 for approximately the same flash rate.

Figure 2.21 shows another variation of the 1.5 V flasher; here, the internal timing resistors are shunted by an external 1k0 resistor, thereby reducing the charge time constant of the circuit and causing the flash rate to increase (to 2.6

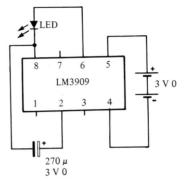

Figure 2.20   *1 Hz flasher consumes an average of 0.77 mA from a 3 V battery*

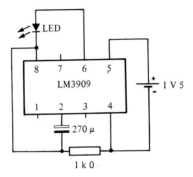

Figure 2.21   *Fast 1V5 blinker. Flash rate is 2.6 Hz and drain current is 1.2 mA*

Hz) and the duty cycle to decrease and the mean current consumption to rise. The circuit gives a far more noticeable flasher indication than the three previous designs, but at the cost of 1.2 mA of current drain.

If you enjoy experimenting with circuits, you can build the variable-rate flasher of *Figure 2.22*; the rate is variable from zero to 20 Hz via the 2k7 pot. The two external 68 R resistors are used to stabilize the duty cycle of the circuit and maintain a fairly steady apparent brilliance level as the rate is varied.

The *Figure 2.23* circuit gives apparently continuous illumination of the LED when powered from a 1.5 V cell. The circuit in fact acts as a 1 kHz square wave generator, the two external 68 R resistors being used to approximately equalize the ON and OFF times of the generator. The circuit gives a fairly dim illumination and has a battery drain of about 4 mA. LED brilliance can be increased by using the alternative connections of *Figure 2.24*, but at the expense of 12 mA of battery drain.

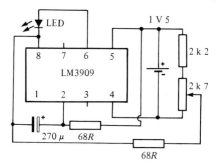

Figure 2.22 *Variable-rate flasher*

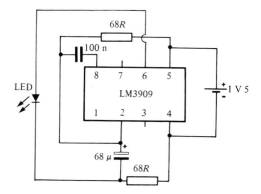

Figure 2.23 *High-efficiency continuous LED indicator operates from 1.5 V*

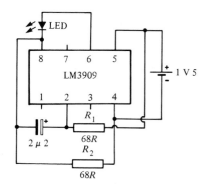

Figure 2.24 *This 1V5 circuit gives an apparently continuous LED indication. Battery drain is 12 mA*

32   LED display circuits

All of the LED flasher circuits that we have looked at in *Figures 2.18* to *2.24* are meant for operation from 1.5 to 3 V supplies. Most of these designs can in fact be used (in slightly modified form) at voltages up to 6V0, as shown in the circuit of *Figure 2.25*. Note in this case that a 68*R* resistor is wired in series with the LED, to limit its drive current to a safe value.

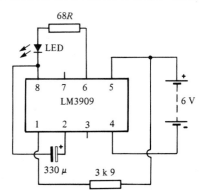

Figure 2.25   *This 6 V flasher operates at about 1 Hz*

The LM3909 IC has a 6V5 zener built in between pins 2 and 4 (not shown in *Figure 2.17*). This fact can be put to practical use in the flasher circuit of *Figure 2.26*, which can be powered from any DC supply in the 85 to 200 volt range;

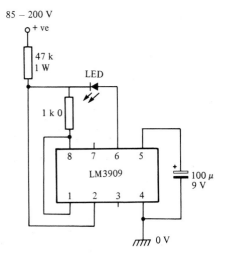

Figure 2.26   *This LED flasher circuit can operate from any supply in the 85 to 200 V range*

the 100 μF timing capacitor is connected between pins 4 and 5 in this application. In some applications it is useful to have several LEDs flashing on and off simultaneously. Four LEDs can, for example, be used to mark the outline of an emergency switch or a first-aid cabinet (often, in emergency situations, mains power services are cut). *Figure 2.27* shows the practical circuit of such a four-LED flasher; note that a 39*R* resistor is wired in series with each LED, with one end connected directly to pin 5 of the IC (the supply positive pin). The circuit has a flash rate of about 1.5 Hz, and draws a mean current of about 1.5 mA from a 1.5 V battery.

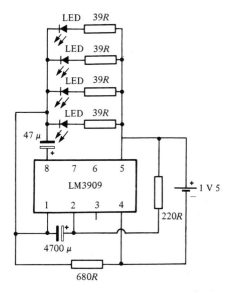

Figure 2.27   *This four-LED flasher operates at 1.5 Hz and drains 1.5 mA from its 1.5 V battery*

Chasers and sequencers

The so-called chaser or sequencer is one of the most popular types of LED display ciruit; it consists of an IC or other electronic device arranged to drive an array of LEDs in such a way that individual LEDs (or small groups of LEDs) turn on and off in a predetermined and repeating sequence, thus producing a visually attractive display in which a ripple of light seems to run along a chain. The 4017B CMOS IC is probably the most popular solid-state device used in chaser/sequencer applications.

4017B basics

The 4017B is actually a CMOS decade counter/divider IC with ten fully decoded outputs that can each be used to directly drive a LED display. If desired, various outputs can be coupled back to the IC control terminals to make the device count to (or divide by) any number from 2 to 9 and then either stop or recycle. Numbers of 4017B ICs can be cascaded to give either multidecade division or to make counters with any desired number of decoded outputs. The 4017B is thus an exceptionally versatile device that can easily be used to chase or sequence a LED display of virtually any desired length.

Figure 2.28 shows the outline, pin designations, and the functional diagram of the 4017B, and Figure 2.29 shows the basic timing diagrams of the device, which incorporates a five-stage Johnson counter and features CLOCK, RESET and CLOCK INHIBIT input terminals. The internal counters are advanced one count at each positive transition of the input clock signal when

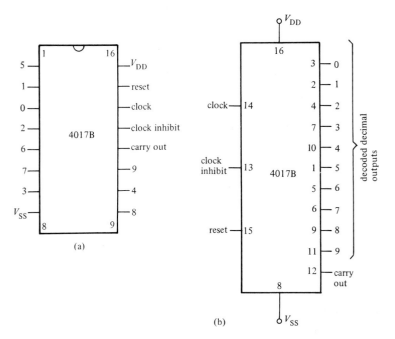

- clock is inhibited by a high signal on pin 13
- counter is reset by a high signal on pin 15
- counter advances on positive transition of clock input

Figure 2.28(a)   *Outline and pin designations of the (b) Functional diagram and data for the 4017B*

the CLOCK INHIBIT and RESET terminals are low. Nine of the ten decoded outputs are low, with the remaining output high, at any given time. The outputs go high sequentially, in step with the clock signal, with the selected output remaining high for one full clock cycle. An additional CARRY OUT signal completes one cycle for every ten clock input cycles, and can be used to ripple-clock additional 4017B ICs in multidecade counting applications.

The 4017B counting cycle can be inhibited by setting the CLOCK INHIBIT terminal high. A high signal on the RESET terminal clears the counter to zero and sets the decoded 0 output terminal high.

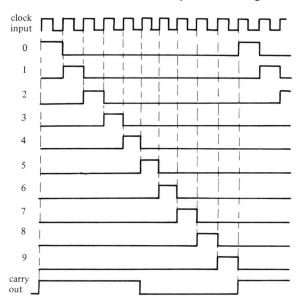

Figure 2.29 *Waveform timing diagram of the 4017B, with its RESET and CLOCK INHIBIT terminals grounded*

4017B chaser/sequencer circuits

Figure 2.30 shows the practical circuit of a 4017B ten-LED chaser, in which IC_1 (a 555 timer chip wired in the astable mode) is used as a variable-speed (via RV_1) clock generator, and the 4017B is wired into the decade counter mode by grounding its CLOCK INHIBIT (pin 13) and RESET (pin 15) control terminals.

The action of the *Figure 2.30* circuit is such that the visual display appears as a moving dot which repeatedly sweeps from the left (LED_0) to the right (LED_9) in ten discrete steps as the 4017B outputs sequentially go high and drive the LEDs on. The LEDs do not, or course, have to be connected in a straight line; they can, for example, be arranged in a circle, in which case the circle will seem to rotate.

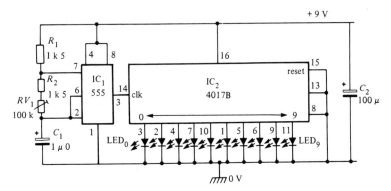

Figure 2.30   *Ten-LED chaser or sequencer can be used with supply voltages up to only 9 V and produces a moving dot display*

Note that the LEDs in the *Figure 2.30* circuit are not provided with current-limiting resistors, and that this circuit can be safely used with maximum supply values of only 9 V. The decoded outputs of the 4017B provide inherent current-limiting under short circuit conditions and, although the manufac-

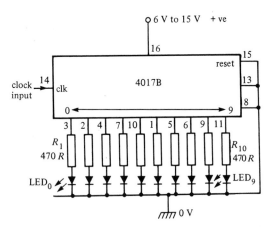

Figure 2.31   *This version of the ten-LED chaser can be used with any supply up to 15 V*

turers do not quote maximum short circuit current-limiting values, practical experience shows that currents of 10 to 15 mA are commonly available from each output of the IC. A maximum device dissipation-per-output-stage figure of 100 mW is quoted on some data sheets, indicating that a volt drop of up to 7 V can be safely developed across a 4017B output stage under the maximum-current condition.

Thus, the *Figure 2.30* LED chaser circuit (which has each LED connected directly between an output and ground) can be safely used at maximum supply values up to only 9 V (allowing for a 2 V drop across each ON LED). At values greater than 9 V the *Figure 2.31* circuit, which has a current-limiting resistor wired in series with each LED, should be used. Note that the main purpose of these resistors is that of reducing the power dissipation of the 4017B.

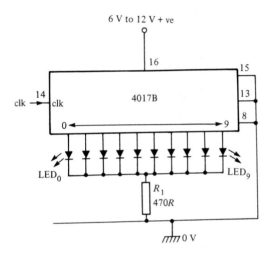

Figure 2.32   *This version of the chaser can be used with supplies up to 12 V maximum*

Figure 2.32 shows a variant that is sometimes used and which can be used with reasonable confidence at supply values up to 12 V maximum. *Figure 2.33* shows a possible equivalent of this circuit when it is powered from a 15 V supply, and illustrates the defect of the design. The action of the 4017B is such that, when a given LED is on, the anodes of all other LEDs are effectively grounded; R_1 thus causes the OFF LEDs to be reverse biased. Because of the low reverse-voltage ratings of LEDs, it will often be found that one of the OFF LEDs will zener at about 5 V, giving the results shown in the diagram and possibly causing a destructive power overload in one of the 4017B output stages.

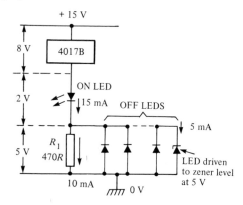

Figure 2.33   *Possible equivalent of the* Figure 2.32 *circuit when powered from a 15 V supply*

Thus, when the 4017B is used to drive LED displays in the 'moving dot' mode, the LEDs can be connected directly to the IC outputs if supply values are limited to 9 V maximum, but at supply values greater than 9 V the LEDs must be connected to the IC outputs via current-limiting resistors. A variety of alternative types of 4017B LED display circuits are shown in *Figures 2.34* to *2.41*.

Alternative LED displays

The output stages of the 4017B can source or sink currents with equal ease. *Figure 2.34* shows how the IC can be used in the sink mode to make a moving hole display in which nine of the ten LEDs are on at any given time, with single LEDs turning off sequentially; if the LEDs are wired in the form of a circle, the circle will seem to rotate. Note that, since all LEDs except one are on at any given time, all LEDs must be provided with current-limiting resistors, to keep the IC power dissipation within safe limits.

In practice, moving dot displays are far more popular than moving hole types. If desired, moving dot displays of the *Figure 2.30* type can be used with fewer than ten LEDs by simply omitting the unwanted LEDs, but in this case the dot will seem to move intermittently, or to scan, since the IC takes ten clock steps to completely sequence and all LEDs will thus be off during the unwanted steps.

If a continuously-moving less-than-ten-LED display is wanted it can be obtained by wiring the first unused output terminal of the 4017B to its pin 15 RESET terminal, as shown, for example, in the four-LED circuit of *Figure*

2.35. Alternatively, the circuit can be made to give an intermittent display with a controlled number of OFF steps by simply taking the appropriate one of the unwanted outputs to the pin 15 RESET terminal. In *Figure 2.36*, for example, the LEDs display for four steps and then blank for four steps, after which the sequence repeats.

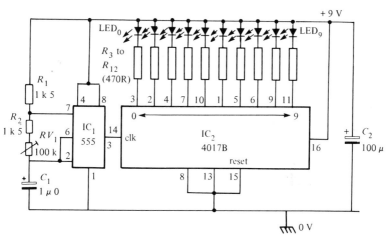

Figure 2.34 *Ten-LED moving hole display*

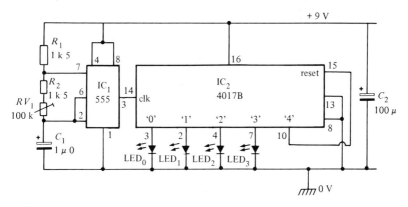

Figure 2.35 *Four-LED continuous moving dot display*

Figure 2.37 shows a rather unusual and very attractive four-LED five-step sequencer, in which all four LEDs are initially on but then turn off one at a time until eventually (in the fifth step) all four LEDs are off: sequencing

40   LED display circuits

details are given in the table in *Figure 2.37*. Note in this circuit that the LEDs are effectively wired in series and that the basic circuit cannot be used to drive more than four LEDs.

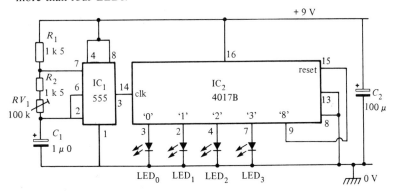

Figure 2.36   *Four-LED intermittent moving dot display with 50% blank period*

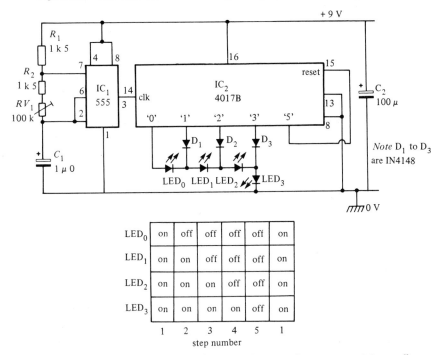

LED$_0$	on	off	off	off	off	on
LED$_1$	on	on	off	off	off	on
LED$_2$	on	on	on	off	off	on
LED$_3$	on	on	on	on	off	on
	1	2	3	4	5	1

step number

Figure 2.37   *Circuit and performance table of a four-LED five-step sequential turn-off display*

Figure 2.38 shows another unusual and attractive LED display. In this case the 4017B runs through a ten-step sequence, with LED_1 on for steps 0 to 3, LED_2 on for steps 4 to 6, LED_3 for steps 7 to 8, and LED_4 on for step 9. The consequence of this action is that the visual display seems to accelerate from LED_1 to LED_4, rather than sweep smoothly from one LED to the next. The acceleration action repeats in each switching cycle, and the cycles repeat ad infinitum.

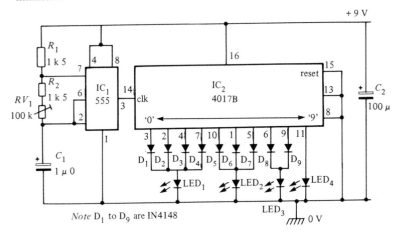

Figure 2.38   *Four-LED continuous accelerator display in which the pattern seems to accelerate from left to right*

Multiplexing

Figure 2.39 shows how the above circuit can be modified to give an intermittent display in which the visual acceleration action occurs for ten clock cycles, but all LEDs then blank for the next twenty cycles, after which the action repeats. The action is as follows.

The 4017B has a CARRY OUT terminal at pin 12. When the IC is used in the divide-by-ten mode (as in *Figures 2.38* and *2.39*) the CARRY OUT terminal produces one output cycle each time the IC completes a decade count. In *Figure 2.39* this signal is used to clock a second 4017B (IC_3), which is wired in the divide-by-three mode with its 0 output fed to gate transistor Q_1. Consequently, for the first ten cycles of a sequence the 0 output of IC_3 is high and Q_1 is biased on, so IC_2 acts in the manner already described for *Figure 2.38*, with its LEDs turning on sequentially and passing current to ground via Q_1. After the tenth clock pulse, however, the 0 output of IC_3 goes low and

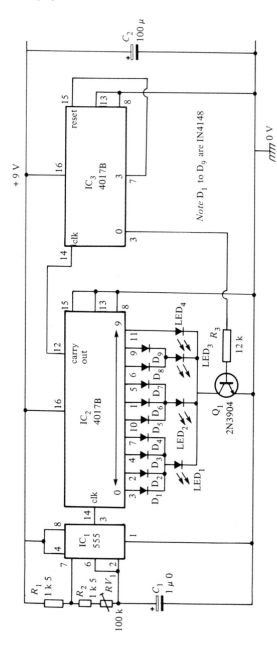

Figure 2.39   *Four-LED intermittent accelerator display, in which acceleration occurs for ten clock steps in every 30*

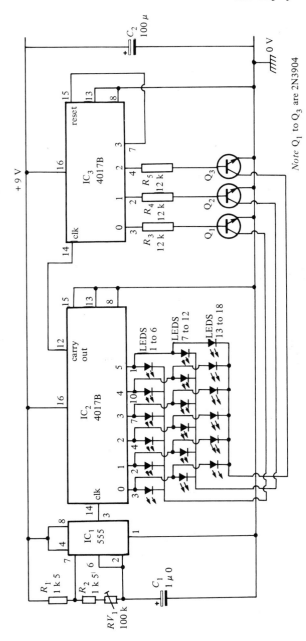

Figure 2.40 *Multiplexed six-LED × three-line moving-dot display. The dot moves intermittently along the lines*

turns Q_1 off, so the LEDs no longer illuminate even though IC_2 continues to sequence. Eventually, after the thirtieth clock pulse, the 0 output of IC_3 again goes high and turns Q_1 on, enabling the display action to repeat again, and so on.

The *Figure 2.39* circuit is a simple example of display multiplexing, in which IC_3 and Q_1 are used to enable or disable a bank of LEDs. *Figure 2.40* shows another example of a multiplexing circuit, in which three lines of six intermittently-sequenced LEDs are sequentially enabled via IC_3 and individual gating transistors, with only one line enabled at any one time. This basic circuit can in fact be expanded to control a ten-line (100 LEDs) matrix display.

Finally, to complete this look at 4017B-based LED sequencer circuits, *Figure 2.41* shows the circuit of a four-bank five-step twenty-LED chaser. Note here that a bank of four LEDs is wired in series in each of the five used outputs of the 4017B, so that four LEDs are illuminated at any given time. In practice, roughly 2 V are dropped across each ON LED, giving a total drop of about 8 V across each ON bank, so the circuit supply voltage must be greater than this value for the circuit to operate. A greater number of LEDs can be used in each bank if the supply voltage value is suitably increased.

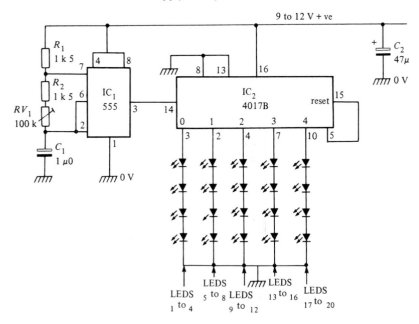

Figure 2.41   *This four-bank five-step twenty-LED chaser must be used with a supply voltage greater than 9 V*

3 LED graph circuits

In Chapter 2 we took a detailed look at LED principles and at a number of practical LED flasher and chaser visual display circuits. In this chapter we continue the LED theme by looking at a variety of LED dot-graph and bar-graph analogue-value display circuits.

LED graph displays

One of the most popular types of multi-LED indicator circuits is the so called analogue-value indicator or graph display, which is designed to drive a chain of linearly-spaced LEDs in such a way that the length of chain that is illuminated is proportional to the analogue value of a voltage applied to the input of the LED-driver circuit, e.g., so that the circuit acts like an analogue voltmeter.

Figure 3.1   *Bar-graph indication of 7 V on a 10 V ten-LED scale*

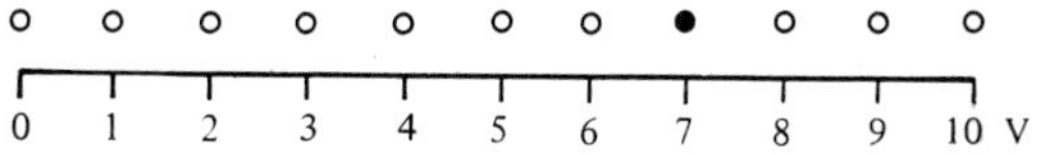

Figure 3.2   *Dot indication of 7 V on a 10 V ten-LED scale*

Practical graph circuits may be designed to generate either a bar-graph display as shown in *Figure 3.1*, or a dot display as in *Figure 3.2*. In a bar-graph display the input voltage value is indicated by the total number of LEDs that

45

are illuminated. In the dot display the input value is indicated by the relative position of a single illuminated LED.

A number of special ICs are available for operating LED analogue-value display systems. The best known of these are the U237 (etc.) family from AEG, and the LM3914 (etc.) family from National Semiconductors. The U237 family are simple dedicated devices which can usefully be cascaded to drive a maximum of ten LEDs in bar mode only. The LM3914 family are more complex and highly versatile devices, which can easily be cascaded to drive as many as 100 LEDs, and can drive them in either bar or dot mode. Both types are listed as bar-graph driver ICs.

IC-driven bar-graph displays make inexpensive and superior alternatives to analogue-indicating moving-coil meters. They are immune to sticking problems, are fast acting, and are unaffected by vibration or by physical attitude. Their scales can easily be given any desired shape. In a given display, individual LED colours can be mixed to emphasize particular sections of the display, and over-range detectors can easily be activated from the driver ICs and used to sound an alarm and/or flash the entire display under the over-range condition.

LED graph displays also have better linearity than conventional moving-coil meters, typical linear accuracy being 0.5%. Scale definition depends on the number of LEDs used; a ten-LED display gives adequate resolution for most practical purposes. A wide range of multi-LED U237-based and LM3914-based graph display circuits are shown in this chapter.

U237 basics

The AEG U237 family of bar-graph driver ICs are simple dedicated devices, housed in eight-pin DIL packages and each capable of directly driving up to five LEDs. The family comprises four individual devices. The U237B and U247B produce a linear-scaled display and are intended to be used as a pair, driving a total of ten LEDs. The U257B and U267B produce a log-scaled display, and are also meant to be used as a pair driving a total of ten LEDs.

All ICs of the U237 family use the same basic internal circuitry, which is shown in block diagram form (together with internal connections) in *Figure 3.3*. The IC houses five sets of Schmitt voltage-comparator/transistor-switches, each of which has its threshold switching or step voltage individually determined by a tapping point on the R_1 to R_6 divider, which is powered from a built-in voltage regulator; the input of each comparator is connected to the pin 7 input terminal of the IC. The IC also houses a constant-current generator (20 mA nominal), and the five external LEDs are wired in series between this generator and ground (pin 1), as shown in *Figure 3.3*.

The basic circuit action is such that groups of LEDs are turned on or off by activating individual switching transistors within the IC. Thus, if Q_3 is turned on it sinks the 20 mA constant-current via LED_1 and LED_2, so $LEDs_1$ and $_2$ turn on and $LEDs_3$ to $_5$ turn off.

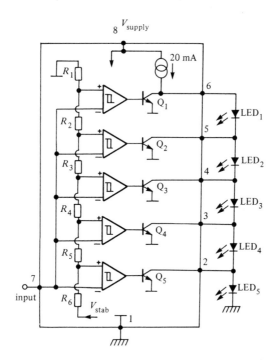

Figure 3.3 *Block diagram of the U237-type bar-graph driver, with basic external connections*

The U237 has step voltages spaced at 200 mV intervals, and *Figure 3.4* shows the states of its five internal transistors at various values of input voltage. Thus, at zero volts input all five transistors are switched on, so Q_1 sinks the full 20 mA of constant current and all five LEDs are off. At 200 mV input Q_1 turns off but all other transistors are on, so Q_2 sinks the 20 mA via LED_1, driving LED_1 on and causing all other LEDs to turn off, and so on. Eventually, at 1 V input, all transistors are off and the 20 mA flows to ground via all LEDs, so all five LEDs are on. Note that the operating current of the circuit is virtually independent of the number of LEDs turned on, so the IC generates negligible RFI as it switches transistors/LEDs.

The four ICs in the U237 family differ only in their values of step voltages, which are determined by the R_1 to R_6 potential divider values. The table in

Figure 3.5 shows the step values of the four ICs. Note that the U237B and U247B are linearly scaled, and can be coupled together to make a ten-LED linear meter with a basic full-scale value of 1 V. The U257B and U267B are log scaled, and can be coupled together to make a ten-LED log meter with a basic full-scale value of 2 V or +6 dB.

V_{IN}	Q_1	Q_2	Q_3	Q_4	Q_5
1.0 V	off	off	off	off	off
0.8 V	off	off	off	off	on
0.6 V	off	off	off	on	on
0.4 V	off	off	on	on	on
0.2 V	off	on	on	on	on
0 V	on	on	on	on	on

Figure 3.4   *States of the U237 internal transistors at various input voltages*

device	step 1	step 2	step 3	step 4	step 5
U237B	200 mV	400 mV	600 mV	800 mV	1.00 V
U247B	100 mV	300 mV	500 mV	700 mV	900 mV
U257B	0.18 V / −15 dB	0.53 V / −6 dB	0.84 V / −1.5 dB	1.19 V / +1.5 dB	2.0 V / +6 dB
U267B	0.1 V / −20 dB	0.32 V / −10 dB	0.71 V / −3 dB	1.0 V / 0 dB	1.41 V / +3 dB

Figure 3.5   *Step-voltage values of the U237 family of bar-graph driver ICs*

Figure 3.6 shows the basic specifications of the U237 family of ICs. Note that the supply voltage range is specified as 8 to 25 V. In practice, the minimum supply voltage is one of the few design points that must be considered when using these ICs, and must at least equal the sum of the ON voltages of the five LEDs, plus a couple of volts to allow correct operation of internal circuitry. Thus, when driving five red LEDs, each with a forward volt drop of 2 V 0, the supply value must be at least 12 V. Different coloured LEDs (with different forward volt drops) can be used together in the circuit, provided that the supply voltage is adequate.

Another usage point concerns the input impedance of the IC. Although the input impedance is typically 100 k, the IC in fact tends to become unstable if fed from a source impedance greater than about 20 k. Ideally, the signal feeding the input should have a source impedance less than 10 k; if the source impedance is greater than 10 k, stability can be enhanced by wiring a 10 n capacitor between pin 7 and pin 1.

parameter	minimum	type	maximum
supply voltage	8 V	12 V	25 V
input voltage			5 V
input current			0.5 mA
maximum supply current		25 mA	30 mA
power dissipation (at 60°C)			690 mW
step tolerance	–30 mV		+30 mV
step hysteresis		5 mV	10 mV
input resistance		100 k	
output saturation voltage			1 V 0

Figure 3.6   *General specifications of the U237 family of ICs*

Practical U237 circuits

Figures 3.7 to *3.12* show some practical ways of using the U237 family of devices. In all these diagrams the supply voltage is shown as being + 12 V to + 25 V, but the reader should keep in mind the constraints already mentioned.

Figure 3.7 shows the practical connections for making a 0–1 V five-LED linear-scaled meter, using a single U237B IC, and *Figure 3.8* shows how a U237B/U247B pair of ICs can be coupled together to make a 0–1 V ten-LED linear-scaled meter. Note in the latter case that the two ICs are operated as

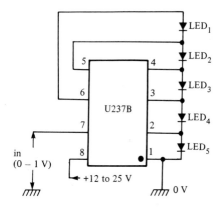

Figure 3.7   *Practical connections for making a 0–1 V five-LED linear-scaled meter*

individual *Figure 3.7* circuits (needing only a five-LED supply voltage), but have their input terminals tied together and have their LEDs *physically* alternated, to give a ten-LED display.

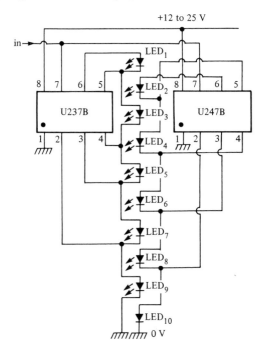

Figure 3.8   *Connections for making a 0–1 V ten-LED linear-scaled meter*

Figure 3.9 shows how the full-scale sensitivity of the basic circuit can be reduced by feeding the input signal to the IC via the R_1 and R_2–RV_1 potential divider, which has a 15:1 ratio and thus gives a full-scale sensitivity of 15 V.

Figures 3.10 and *3.11* show how the basic *Figure 3.7* circuit can be used to indicate the value of a physical parameter (such as light, heat, etc.) that can be represented by the analogue resistance value of transducer R_T. In both cases, the transducer is effectively fed from a constant-current source, so that the input voltage to the IC is directly proportional to the transducer resistance value.

In *Figure 3.10* the transducer current is derived from the regulated supply line via R_1–RV_1, and current constancy relies on the fact that the regulated supply voltage is large relative to the 1 V full-scale sensitivity of the meter. In *Figure 3.11*, current constancy is ensured by the ZD_1–Q_1–etc. constant-current generator.

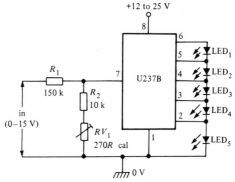

Figure 3.9   *Method of reducing the sensitivity of the* Figure 3.7 *circuit, to make a 0–15 V five-LED meter*

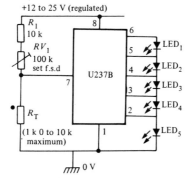

Figure 3.10   *Simple method of using a transducer sensor to indicate the value of a physical quantity*

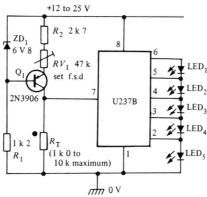

Figure 3.11   *Alternative method of using a transducer sensor to indicate the value of a physical quantity*

Finally, *Figure 3.12* shows how the U267 log IC can be used to make a five-LED audio-level meter. A ten-LED meter can be made by connecting the R_1–R_2–R_3–C_1–D_1 input circuit to the input of a U257B/V267B pair of ICs wired in the configuration shows in *Figure 3.8*.

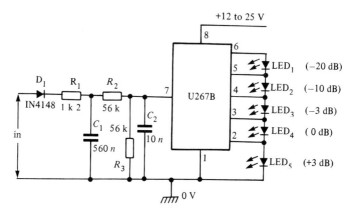

Figure 3.12   *Five-LED AF-level meter. A ten-LED version can be made by using a U257B/U267B pair of ICs*

LM3914 family basics

The LM3914 family of dot-/bar-graph driver ICs are manufactured by National Semiconductors. They are fairly complex but highly versatile devices, housed in eighteen-pin DIL packages and each capable of directly driving up to ten LEDs in either the dot or the bar mode. The family comprises three devices, these being the LM3914, the LM3915 and the LM3916; they all use the same BASIC internal circuitry (see *Figure 3.13*), but differ in the style of scaling of the LED-driving output circuitry, as shown in *Figure 3.14*.

Thus, the LM3914 is a linearly-scaled unit, specifically intended for use in LED voltmeter applications in which the number of illuminated LEDs gives a direct indication of the value of input volts. The LM3915, on the other hand, has a log-scaled output designed to span 0 to -27 dB in ten -3 dB steps, and is specifically designed for use in power meter applications, etc. Finally, the LM3916 has a semi-log scale, and is specifically designed for use in VU meter applications.

All three devices of the LM3914 family use the same basic internal circuitry, and *Figure 3.13* shows the specific internal circuit of the linear-scaled LM3914, together with the connections for making it act as a simple ten-LED

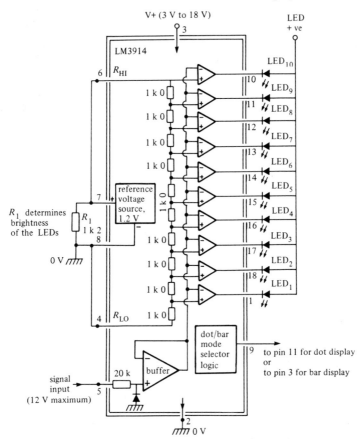

Figure 3.13   *Internal circuit of the LM3914, with connections for making a ten-LED, 0–1.2 V, linear meter with dot or bar display*

0 to 1V2 meter. The IC contains ten voltage comparators, each with its non-inverting terminal taken to a specific tap on a 'floating' precision multistage potential divider and with all inverting terminals wired in parallel and taken to input pin 5 via a unity gain buffer amplifier. The output of each comparator is externally available, and can sink up to 30 mA; the sink currents are internally limited, and can be externally pre-set via a single resistor (R_1).

The IC also contains a floating 1.2 V reference source between pins 7 and 8. In *Figure 3.13* the reference is shown externally connected to the internal (pins 4 to 6) potential divider. Note that pins 8 and 4 are shown grounded, so in this case the bottom of the divider is at zero volts and the top is at 1.2 V. The IC also contains a logic network that can be externally set to give either a dot or a bar display from the outputs of the ten comparators. The IC operates as follows.

Assume that the IC logic is set for bar-mode operation, and that the 1.2 V reference is applied across the internal ten-stage divider as shown. Thus, 0.12 V is applied to the inverting or reference input of the lower comparator, 0.24 V to the next, 0.36 V to the next, and so on. If a slowly rising input voltage is now applied to pin 5 of the IC the following sequence of actions takes place.

LED number	typical threshold – point value, at 10 V f.s.d.					
	LM3914	LM3915			LM3916	
	V	V	dB	V	dB	VU
1	1.00	.447	−27	.708	−23	−20
2	2.00	.631	−24	2.239	−13	−10
3	3.00	.891	−21	3.162	−10	−7
4	4.00	1.259	−18	3.981	−8	−5
5	5.00	1.778	−15	5.012	−6	−3
6	6.00	2.512	−12	6.310	−4	−1
7	7.00	3.548	−9	7.079	−3	0
8	8.00	5.012	−6	7.943	−2	+1
9	9.00	7.079	−3	8.913	−1	+2
10	10.00	10.00	0	10.00	0	+3

Figure 3.14   *Threshold-point values of the LM3914/15/16 range of ICs, when designed to drive ten-LEDs at a full-scale sensitivity of 10 V*

When the input voltage is zero the outputs of all ten comparators are disabled and all LEDs are off. When the input voltage reaches the 0.12 V reference value of the first comparator its output conducts and turns LED_1 on. When the input reaches the 0.24 V reference value of the second comparator its output also conducts and turns on LED_2, so at this stage LED_1 and LED_2 are both on. As the input voltage is further increased progressively more and more comparators and LEDs are turned on until eventually, when the input rises to 1.2 V, the last comparator and LED_{10} turn on, at which point all LEDs are on.

A similar kind of action is obtained when the LM3914 logic is set for dot mode operation, except that only one LED is on at any given time; at 0 V no LEDs are on, and at 1.2 V and greater only LED_{10} is on.

Some finer details

In *Figure 3.13*, R_1 is shown connected between pins 7 and 8 (the output of the 1.2 V reference), and determines the ON currents of the LEDs. The ON current of each LED is roughly ten times the output current of the 1.2 V source, which can supply up to 3 mA and thus enables LED currents of up to 30 mA to be set via R_1. If, for example, a total resistance of 1k2 (equal to the paralleled values of R_1 and the 10 k of the ICs internal potential divider) is placed across pins 7 and 8 the 1.2 V source will pass 1 mA and each LED will pass 10 mA in the ON mode.

Note from the above that the IC can pass total currents up to 300 mA when used in the bar mode with all ten LEDs on. The IC has a maximum power rating of only 660 mW, so there is a danger of exceeding this rating when the IC is used in the bar mode. In practice, the IC can be powered from DC supplies in the range 3 to 25 V, and the LEDs can use the same supply as the IC or can be independently powered; this latter option can be used to keep power dissipation of the IC at minimal level.

The internal ten-stage potential divider of the IC is floating, with both ends externally available for maximum versatility, and can be powered from either the internal reference or from an external source or sources. If, for example, the top of the chain is connected to a 10 V source, the IC will function as a 0–10 V meter if the low end of the chain is grounded, or as a restricted-range 5–10 V meter if the low end of the chain is connected to a 5 V source. The only constraint on using the divider is that its voltage must not be greater than 2 V less than the ICs supply voltage (which is limited to 25 V maximum). The input (pin 5) to the IC is fully protected against overload voltages up to plus or minus 35 V.

The internal voltage reference of the IC produces a nominal output of 1.28 V (limits are 1.2 V to 1.32 V), but can be externally programmed to produce effective reference values up to 12 V (we show how later).

The IC can be made to give a dot mode display by wiring pin 9 to pin 11, or a bar display by wiring pin 9 to positive-supply pin 3.

Finally, note that the major difference between the three members of the LM3914 family of ICs lays in the values of resistance used in the internal ten-stage potential divider. In the LM3914 all resistors in the chain have equal values, and thus produce a linear display of ten equal steps. In the LM3915 the resistors are logarithmically weighted, and thus produce a log display that spans 30 dB in ten 3 dB steps. In the LM3916 the resistors are weighted in semilog fashion and produce a display that is specifically suited to VU-meter applications.

Let us now move on and look at some practical applications of this series of devices, paying particular attention to the linear LM3914 IC.

Dot-mode voltmeters

Figures 3.15 to *3.19* show various ways of using the LM3914 IC to make ten-LED dot-mode voltmeters with a variety of full-scale deflection (f.s.d.)

sensitivities. Note in all these circuits that pin 9 is wired to pin 11, to give dot-mode operation, and that a 10 μ capacitor is wired directly between pins 2 and 3 to enhance circuit stability.

Figure 3.15 shows the connections for making a variable-range (1.2 V to 1000 V f.s.d.) voltmeter. The low ends of the internal reference and divider are

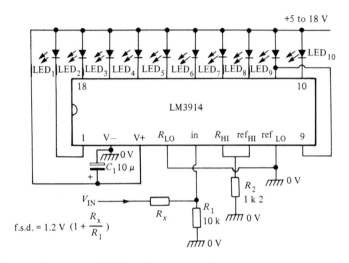

Figure 3.15   *1.2 V to 1000 V f.s.d. dot mode voltmeter*

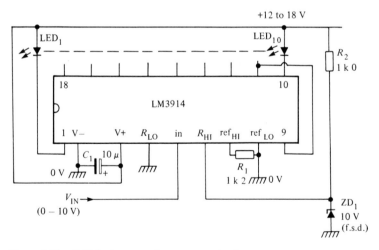

Figure 3.16   *10 V f.s.d. meter using an external reference*

grounded and their top ends are joined together, so the meter has a basic full-scale sensitivity of 1.2 V, but variable ranging is provided by the R_x-R_1 potential divider at the input of the circuit. Thus, when R_x is zero, f.s.d. is 1.2 V, but when R_x is 90 k the f.s.d. is 12 V. Resistor R_2 is wired across the internal reference and sets the ON currents of the LEDs at about 10 mA.

Figure 3.16 shows how to make a fixed-range 0–10 V meter, using an external 10 V zener (connected to the top of the internal divider) to provide a reference voltage. The supply voltage to this circuit must be at least 2 V greater than the zener reference voltage.

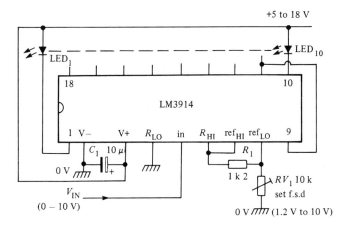

Figure 3.17   *An alternative variable-range (1.2 V to 10 V) dot-mode voltmeter*

Figure 3.17 shows how the internal reference of the IC can be made to effectively provide a variable voltage, enabling the meter f.s.d. value to be set anywhere in the range 1.2 V to 10 V. In this case the 1 mA current (determined by R_1) of the floating 1.2 V internal reference flows to ground via RV_1, and the resulting RV_1-voltage raises the reference pins (pins 7 and 8) above zero. If, for example, RV_1 is set to 2k4, pin 8 will be at 2.4 V and pin 7 at 3.6 V. RV_1 thus enables the pin 7 voltage (connected to the top of the internal divider) to be varied from 1.2 V to about 10 V, and thus sets the f.s.d. value of the meter within these values.

Figure 3.18 shows the connections for making an expanded-scale meter that, for example, reads voltages in the range 10 to 15 V. RV_2 sets the LED current at about 12 mA, but also enables a reference value in the range 0–1.2 V to be set on the low (pin 4) end of the internal divider. Thus, if RV_2 is set to apply 0.8 V to pin 4 the basic meter will read voltages in the range 0.8 to 1.2 V only. By fitting potential divider R_x–RV_1 to the input of the circuit this range can be amplified to, say, 10–15 V, or whatever range is desired.

58   LED graph circuits

Finally, *Figure 3.19* shows an expanded scale dot-mode voltmeter that is specifically designed to indicate the value of a car's battery (12 V nominal). In this case R_2-RV_2 are effectively set to give a basic range of 2.4 to 3.6 V, but the

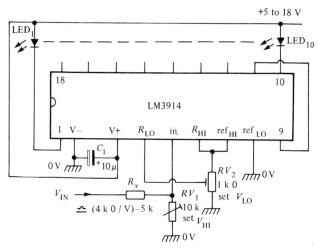

Figure 3.18   *Expanded-scale (10 V–15 V, etc.) dot-mode voltmeter*

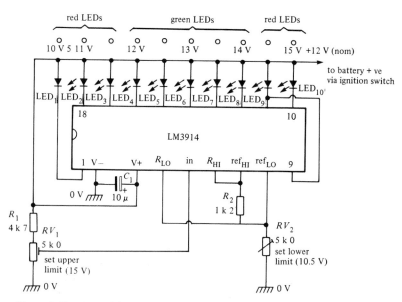

Figure 3.19   *Expanded-scale dot-mode car voltmeter*

input to the circuit is derived from the positive supply rail via the R_1–RV_1 potential divider, and the indicated volts reading thus corresponds to a pre-set multiple of the basic range value. As shown in the diagram, red and green LEDs can be used in the display, arranged so that green LEDs illuminate when the voltage is in the safe range 12 to 14 V.

To calibrate the above circuit, first set the supply to 15 V and adjust RV_1 so that LED_{10} just turns on. Reduce the supply to 10 V and adjust RV_2 so that LED_1 just turns on. Recheck the settings of RV_1 and RV_2. The calibration is then complete and the unit can be installed in the car by taking the 0 volt lead to the chassis and the $+12$ V lead to the car's battery via the ignition switch.

Bar-mode voltmeters

The dot-mode circuits of *Figures 3.15* to *3.19* can be made to give bar-mode operation by simply connecting pin 9 to pin 3, rather than to pin 11. When using the bar mode, however, it must be remembered that the IC's power rating must not be exceeded by allowing excessive output-terminal voltages to be developed when all ten LEDs are on. LEDs drop roughly 2 V when they are conducting, so one way around this problem is to power the LEDs from their own low-voltage (3 to 5 V) supply, as shown in *Figure 3.20*.

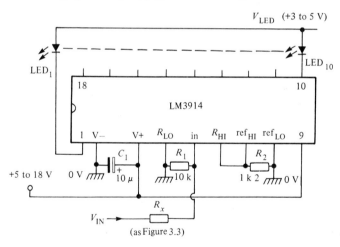

Figure 3.20 *Bar-display voltmeter with separate LED supply*

An alternative solution is to power the IC and the LEDs from the same supply, but to wire a current-limiting resistor in series with each LED, as shown in *Figure 3.21*, so that the IC's output terminal saturates when the LEDs are on.

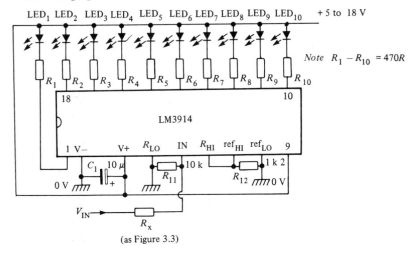

Figure 3.21   *Bar-display voltmeter with common LED/IC supply*

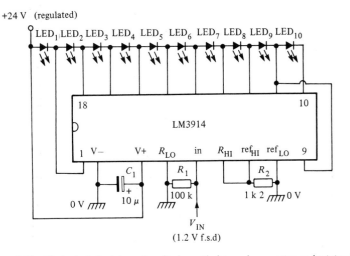

Figure 3.22   *Method of obtaining a bar display with dot-mode operation and minimal current consumption*

Figure 3.22 shows another way of obtaining a bar display without excessive power dissipation. Here, the LEDs are all wired in series, but with each one connected to an individual output of the IC, and the IC is wired for dot-mode operation. Thus, when, for example, LED_5 is on it draws its current via LED_1

to LED$_4$, so all five LEDs are on and the total LED current equals that of a single LED, and total power dissipation is quite low. The LED supply to this circuit must be greater than the sum of all LED volt-drops when all LEDs are on, but must be within the voltage limits of the IC; a regulated 24 V supply is thus needed.

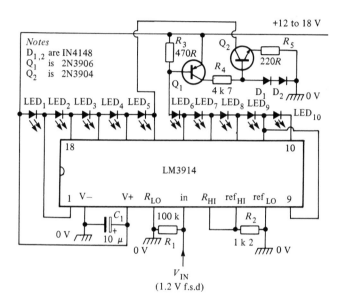

Figure 3.23   *Modification of the* Figure 3.22 *circuit, for operation from unregulated 12 V to 18 V supplies*

Figure 3.23 shows a very useful modification which enables the above circuit to be powered from unregulated supplies within the 12 to 18 V range. In this case the LEDs are split into two chains, and the transistors are used to switch on the lower (LED$_1$ to LED$_5$) chain when the upper chain is active; the maximum total LED current equals twice the current of a single LED.

20-LED voltmeters

Figure 3.24 shows how two LM3914 ICs can be interconnected to make a twenty-LED dot-mode voltmeter. Here, the input terminals of the two ICs are wired in parallel, but IC$_1$ is configured so that it reads 0 to 1.2 V, and IC$_2$ is configured so that it reads 1.2 to 2.4 V. In the latter case the low end of the IC$_2$ potential divider is coupled to the 1.2 V reference of IC$_1$, and the top end of the divider is taken to the top of the 1.2 V reference of IC_2, which is raised 1.2 V above that of IC_1.

The above design is wired for dot-mode operation; note in this case that pin 9 of IC_1 is wired to pin 1 of IC_2, and pin 9 of IC_2 is wired to pin 11 of IC_2. Also note that a 22 k resistor is wired in parallel with LED_9 of IC_1.

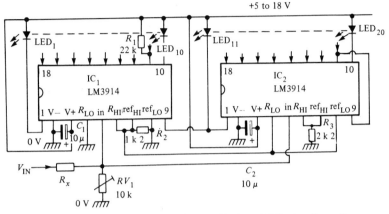

Figure 3.24   *Dot-mode twenty-LED voltmeter (f.s.d. = 2.4 V when $R_x = 0$)*

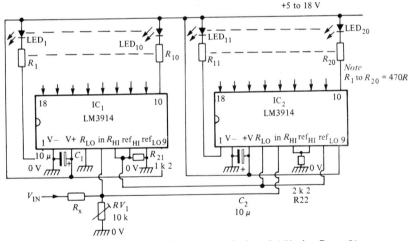

Figure 3.25   *Bar-mode twenty-LED voltmeter (f.s.d. = 2.4 V when $R_x = 0$)*

Figure 3.25 shows the connections for making a twenty-LED bar-mode voltmeter. The connections are similar to those of *Figure 3.24*, except that pin 9 is taken to pin 3 of each IC, and a 470*R* current-limiting resistor is wired in series with each LED to reduce the power dissipation of the ICs.

To conclude this look at LM3914 circuits, *Figure 3.26* shows a simple frequency-to-voltage converter that can be used to convert either of the *Figure 3.24* or *3.25* circuits into twenty-LED tachometers or r.p.m. meters. This converter should be interposed between the vehicle's contact-breaker points and the input of the voltmeter circuit. In *Figure 3.26*, the C_2 value of 22 n is the optimum value for a full-scale range of 10,000 r.p.m. on a four-cylinder four-stroke engine. For substantially lower full-scale r.p.m. values, the C_2 value may have to be increased: the value may have to be reduced on vehicles with six or more cylinders.

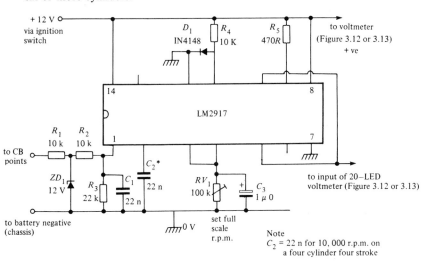

Figure 3.26   *Car tacho conversion circuit for use with a twenty-LED voltmeter* (Figure 3.24 *or* 3.25)

LM3915/LM3916 circuits

The LM3915 log and LM3916 semilog ICs operate in the same basic way as the LM3914, and can in fact be directly substituted in most of the circuits shown in *Figures 3.15* to *3.25*. In most practical applications, however, these particular ICs are used to give a meter indication of the value of an AC input signal, and the simplest way of achieving such a display is to connect the AC signal directly to the pin 5 input terminal of the IC, as shown in *Figure 3.27*. The IC responds only to the positive halves of such input signals, and the number of illuminated LEDs is thus proportional to the instantaneous peak value of the input signal. In such circuits, the IC should be operated in the dot mode and set to give about 30 mA of LED drive current.

The *Figure 3.27* circuit is that of a simple LM3915-based audio power meter. Pin 9 is left open-circuit to ensure dot-mode operation, and R_1 has a value of 390R to give a LED current of about 30 mA. The meter gives audio power indication over the range 200 mW to 100 W.

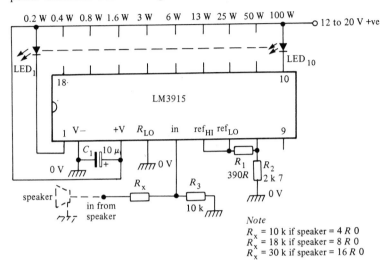

Figure 3.27   *Simple audio power meter*

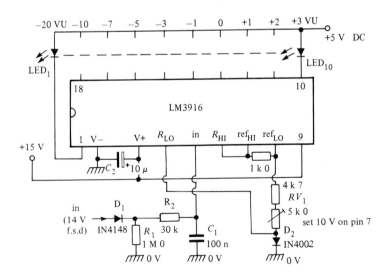

Figure 3.28   *Simple VU-meter*

A more sophisticated way of using these ICs to show the value of an AC input signal is to use a half-wave converter to change the AC signal into DC that is then fed to the input of the IC. *Figures 3.28* and *3.29* show practical LM3916-based VU-meter circuits of this type.

In *Figure 3.28*, the input signal is converted to DC via the simple D_1–R_1–R_2–C_1 network. Note in this case that rectifier D_2 is used to compensate for the forward volt drop of D_1. Also note that this particular circuit operates in the bar mode and uses separate supplies for the IC and the LED display.

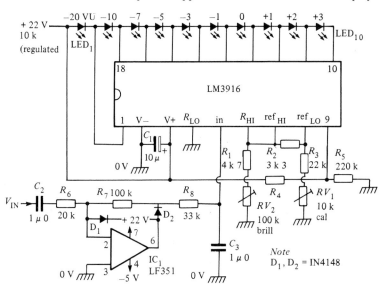

Figure 3.29   *Precision VU-meter, with low current drain*

Finally, to complete this look at the LM3914 range of devices, *Figure 3.29* shows how the LM3916 can be used as a precision VU-meter by using a precision half-wave rectifier (IC_1) to give AC/DC conversion. Note in this circuit that the LEDs are wired in series and IC_2 is wired in the dot mode, to give a low-consumption bar display of the type shown in *Figure 3.22*. To set up this circuit, simply adjust RV_1 to set 10 V on pin 7; RV_2 can then be used as a LED brightness control.

4 Seven-segment displays

In Chapters 2 and 3 we have looked at a variety of optoelectronic display circuits that use simple LEDs as their display elements. In this chapter we continue the display theme by looking at seven-segment alphanumeric displays and at a variety of associated driver devices and circuitry.

Seven-segment displays

A very common requirement in modern electronics is that of displaying alphanumeric characters. Digital watches, pocket calculators, and digital multimeters and frequency meters are all examples of devices that make use of such displays. The best known type of alphanumeric display is the so-called seven-segment display, which comprises seven independently-accessible photoelectric segments (such as LEDs or liquid crystals, or gas-discharge or fluorescent elements, etc.) arranged in the form shown in *Figure 4.1*. The segments are conventionally notated from a to g in the manner shown in

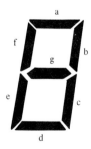

Figure 4.1 *Standard form and notations of a seven-segment display*

66

Figure 4.1, and it is possible to make them display any number (numeral) from 0 to 9 or alphabetic character from A to F (in a mixture of upper and lower case letters) by activating these segments in various combinations, as shown in the truth table of *Figure 4.2*.

segments (✓ = ON)							display	segments (✓ = ON)							display
a	b	c	d	e	f	g		a	b	c	d	e	f	g	
✓	✓	✓	✓	✓	✓		0	✓	✓	✓	✓	✓	✓	✓	8
	✓	✓					1	✓	✓	✓			✓	✓	9
✓	✓		✓	✓		✓	2	✓	✓	✓		✓	✓	✓	A
✓	✓	✓	✓			✓	3			✓	✓	✓	✓	✓	b
	✓	✓			✓	✓	4	✓			✓	✓	✓		C
✓		✓	✓		✓	✓	5		✓	✓	✓	✓		✓	d
✓		✓	✓	✓	✓	✓	6	✓			✓	✓	✓	✓	E
✓	✓	✓					7	✓				✓	✓	✓	F

Figure 4.2   *Truth table for a seven-segment display*

Practical seven-segment display devices must be provided with at least eight external connection terminals; seven of these give access to the individual photoelectric segments, and the eighth provides the essential common connection to all segments. If the display is of the LED type, the seven individual LEDs may be arranged in the form shown in *Figure 4.3*, in which all LED anodes are connected to the common terminal, or they may be arranged as in *Figure 4.4*, in which all LED cathodes are connected to the common

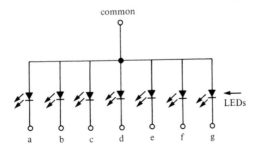

Figure 4.3   *Schematic diagram of a common anode seven-segment LED display*

terminal. In the former case the device is known as a common-anode seven-segment display; in the latter case the device is known as a common-cathode seven-segment display.

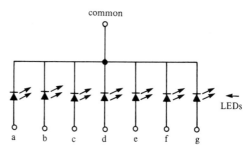

Figure 4.4 *Schematic diagram of a common cathode seven-segment LED display*

Seven-segment display/drivers

In most practical applications, seven-segment displays are used to give a visual indication of the output states of digital ICs such as decade counters and latches, etc. These outputs are usually in four-bit BCD (binary coded decimal) form, and are thus not suitable for directly driving seven-segment displays. Consequently, special BCD-to-seven-segment decoder/driver ICs are available to convert the BCD signal into a form suitable for driving these displays, and are connected between the BCD signals and the display in the manner shown in *Figure 4.5*. The table in *Figure 4.6* shows the relationship between the BCD signals and the displayed seven-segment numerals.

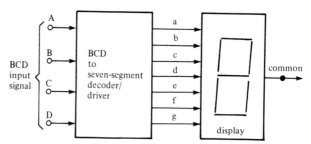

Figure 4.5 *Basic connections of a BCD-to-seven-segment decoder/driver IC*

In practice, BCD-to-seven-segment decoder/driver ICs are usually available in a dedicated form that is suitable for driving only a single class of display unit, e.g., either common-anode LED type, or common-cathode LED

BCD signal				display	BCD signal				display
D	C	B	A		D	C	B	A	
0	0	0	0		0	1	0	1	
0	0	0	1		0	1	1	0	
0	0	1	0		0	1	1	1	
0	0	1	1		1	0	0	0	
0	1	0	0		1	0	0	1	

0 = logic low

1 = logic high

Figure 4.6   Truth table of a BCD-to-seven-segment decoder/driver

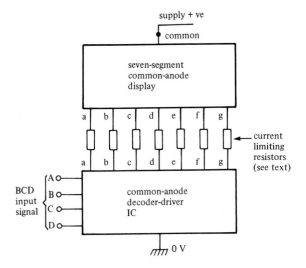

Figure 4.7   Method of driving a common-anode display

type, or liquid crystal displays (LCDs). *Figures 4.7* to *4.9* show the methods of interconnecting each of these IC and display types.

Note in the case of the LED circuits (*Figures 4.7* and *4.8*) that if the IC outputs are unprotected (as in the case of most TTL ICs), a current-limiting resistor must be wired in series with each display segment; most CMOS ICs have internally current-limited outputs, and do not require the use of these external resistors. Finally, note in the case of the *Figure 4.9* LCD-driving circuit that the common or backplane (BP) terminal of the display must be driven from a symmetrical square wave signal derived from the phase output terminal of the IC.

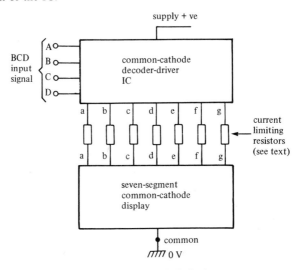

Figure 4.8 *Method of driving a common-cathode display*

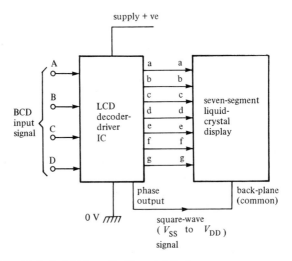

Figure 4.9 *Method of driving a liquid-crystal display*

Cascaded displays

If required, several sets of seven-segment displays and matching decoder/driver ICs can be cascaded and used to make multidecade display systems. *Figure 4.10* shows how three sets of these ICs/displays can be used with a trio of decade counters to make a simple digital-readout frequency meter. Here, the amplified external frequency signal is fed to the input of the series-connected counters via one input of a 2-input AND gate, which has its other (GATE) input waveform derived from a built-in timebase generator. The circuit's operating sequence is as follows:

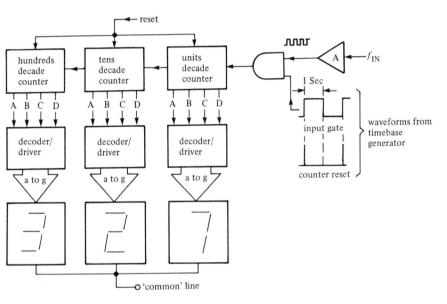

Figure 4.10   *Simple digital frequency meter circuit*

When the timebase GATE input signal is low the AND gate is closed and no input signals reach the counters. At the moment that the timebase GATE signal switches high a brief RESET pulse is fed to all three counters, setting them all to zero count; simultaneously, the input gate opens, and remains open for a period of precisely one second, during which time the input-frequency pulses are summed by the counters. At the end of the one second period the gate closes and the timebase GATE signal goes low again, thus ending the count and enabling the displays to give a steady reading of the total pulse count (and hence frequency). The whole process then repeats again one second later, when the timebase GATE signal again goes high.

Display latching

The simple cascaded system described above sufers from a major defect, in that the display becomes a blur during the actual counting period, becoming stable and readable only when each count is completed and the input gate is closed. This 'blur-and-read' type of display is very annoying to watch. *Figure 4.11* shows an improved frequency meter circuit that uses display latching to overcome the above defect. Here, a four-bit data latch is wired between the output of each counter and the input of its decoder/driver IC. This circuit operates as follows.

At the moment that the timebase GATE signal goes high a RESET pulse is fed to all counters, setting them to zero. Simultaneously, the input gate is opened and the counters start to sum the input signal pulses. This count continues for precisely one second, and during this period the four-bit latches prevent the counter output signals from reaching the display drivers; the display thus remains stable during this period. At the end of the one second count period the AND gate closes and terminates the count, and simultaneously a brief LATCH ENABLE pulse is fed to all latches, causing the prevailing BCD outputs of each counter to be latched into memory and

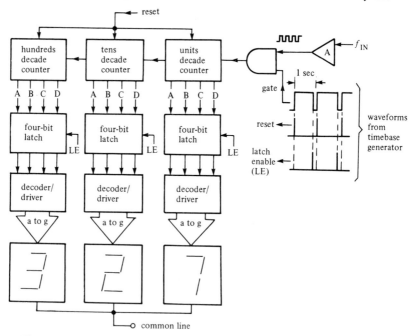

Figure 4.11 *Improved digital frequency meter circuit*

thence fed to the display via the decoder/driver ICs, thus causing the display to give a steady reading of the total pulse count (and thus the input frequency). A few moments later the sequence repeats again, with the counters resetting and then counting the input frequency pulses for one second, during which time the display gives a steady reading of the results of the previous count, and so on.

The *Figure 4.11* circuit thus generates a stable display that is updated once every second or so; in practice, the actual count period of this and the *Figure 4.10* circuit can be made any decade multiple or submultiple of one second, provided that the output display is suitably scaled. Note that in reality many decoder/driver ICs incorporate built-in four-bit data latches.

Multiplexing

Note from the *Figure 4.10* and *4.11* circuits that a total of at least twenty-one connections must be made between the IC circuitry and the seven-segment displays of a three-digit readout unit; a total of at least seventy connections are needed if a ten-digit display is used. In practice, the number of IC-to-display connections can be greatly reduced by using the technique known as multiplexing. This technique can be understood with the aid of *Figures 4.12* and *4.13*.

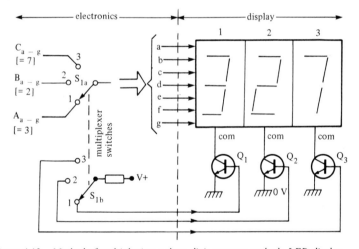

Figure 4.12   *Method of multiplexing a three-digit common-cathode LED display*

Figure 4.12 shows how each digit of a three-digit common-cathode LED display can be individually activated using a total of only ten external connections; the circuitry to the left of the dotted line should be regarded as 'electronic', and to the right of the line as 'display' circuitry. In the display, all a segments are connected together, as also are all other (b to g) sets of segments, so that a total of only seven external a to g connections are made to the display irrespective of the number of digits used. Note, however, that none of the seven-segment displays are influenced by signals on these segment wires unless a display is enabled by connecting its common terminal to ground, and in *Figure 4.12* this is achieved by activating switching transistors Q_1 to Q_3 via suitable external signals, which require the use of only one additional connection per display digit.

Note in *Figure 4.12* that three different sets of segment data can be selected via switch S_{1a}, which in reality would take the form of a ganged seven-pole three-way electronic switch (with one pole dedicated to each of the seven segment lines), and that any one of the three display digits can be selected via S_{1b} and Q_1 to Q_3. These switches are ganged together and provide the actual multiplexer action, and should be regarded as fast-acting electronic switches that repeatedly switch through positions 1, 2 and 3. The operating sequence of the circuit is as follows.

Assume initially that the switch is in position 1. Under this condition S_{1a} selects segment data Aa–g, and S_{1b} activates display 1 via Q_1, so that display 1 shows the number 3. A few moments later the switch jumps to position 2, selecting segment data Ba–g and activating display 2 via Q_2, so that display 2 shows the number 2. A few moments later the switch jumps to position 3, causing display 3 to show the number 7. A few moments later the whole cycle starts to repeat again, and so on ad infinitum. In practice, dozens of these cycles occur each second, so the eye does not see the displays as being turned on and off individually, but sees each of them as a steady display with the trio apparently showing the steady number 327, or whatever other number is dictated by the segment data.

Note from the above description that, since each display is turned on for only one third of each cycle, the mean current consumption of each display is only one third of the peak display current, and the LED brightness levels are correspondingly reduced. In practical multiplexers the peak display current is made fairly high, to give adequate display brightness.

Figure 4.13 shows an example of an improved multiplexing (MUX) technique, as applied to a three-digit frequency meter. In this case the MUX is interposed between the outputs of the three BCD data latches and the input of the BCD-to-seven-segment decoder/driver IC. This technique has two major advantages. First, it calls for the use of only a single decoder/driver IC, irrespective of the number of readout digits used. Second, it calls for the use of a MUX incorporating only five ganged three-way sequencing switches (one for the control data and four for the BCD data), rather than the eight ganged three-way switches (one for the control data and seven for the segment data) called for in the *Figure 4.12* system.

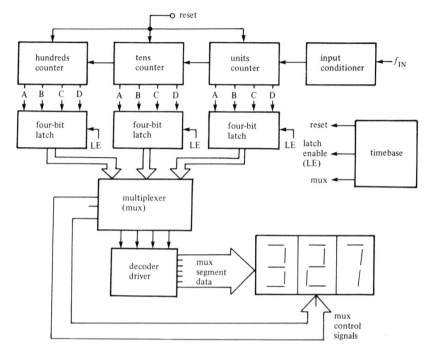

Figure 4.13 *Realistic implementation of the multiplexing technique in a three-digit frequency meter*

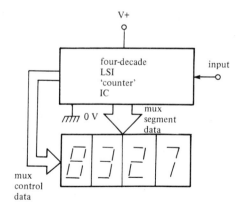

Figure 4.14 *Four-digit counter circuit, using a LSI chip*

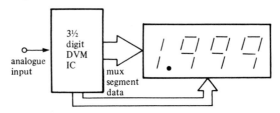

Figure 4.15   *3½-digit DVM using a LSI chip*

In practice, all of the counting, latching, multiplexing, decoding, timing and display-driving circuitry of *Figure 4.13* (and a great deal more) can easily be incorporated in a single LSI (large scale integration) chip that needs only twenty or so pins to make all necessary connections to the power supply, displays, and inputs, etc. Thus, a complete four-digit counter can be implemented using a dedicated IC in a circuit such as that shown in *Figure 4.14*, or a 3½ digit DVM (digital volt meter) can be implemented using a circuit such as that shown in *Figure 4.15*.

Ripple blanking

If the basic four-digit circuit of *Figure 4.14* is used to measure a count of 27 it will actually give a reading of 0027, unless steps are taken to provide automatic suppression of the two (unwanted) leading zeros. Similarly, if the 3½-digit circuit of *Figure 4.15* is used to measure 0.1 V it will actually give a display of 0.100 V, unless steps are taken to provide automatic suppression of the two (unwanted) trailing zeros.

In practice, automatic blanking of leading and/or trailing zeros can be obtained by using a ripple blanking technique, as illustrated in *Figures 4.16* and *4.17*. In these diagrams, each decoder/driver IC has a BCD input and a seven-segment output, and is provided with ripple blanking input (RBI) and output (RBO) terminals. If these terminals are active high they will have the following characteristics.

If the RBI terminal is held low (at logic-0), the seven-segment outputs of the IC are enabled but the RBO terminal is disabled (held low). If the RBI terminal is biased high (at logic-1), the seven-segment outputs become disabled in the presence of a BCD 0000 input (= decimal zero), and the RBO output goes high under the same condition. Thus, the RBO terminal is normally low and goes high only if a BCD 0000 input is present at the same time as the RBI terminal is high. With these characteristics in mind, refer now to *Figures 4.16* and *4.17*.

Figure 4.16 shows the ripple blanking technique used to provide leading-

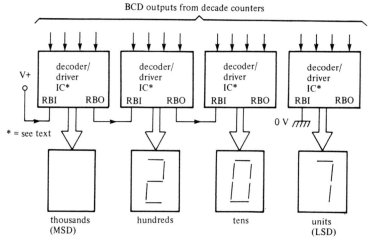

Figure 4.16 *Ripple-blanking used to give leading-zero suppression in a four-digit counter*

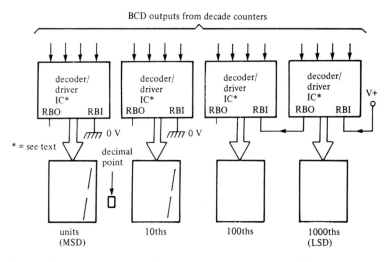

Figure 4.17 *Ripple-blanking used to give trailing-zero suppression of the last two digits of a 3½-digit DVM readout*

zero suppression in a four-digit display that is reading a count of 207. Here, the RBI input of the thousands or most significant digit (MSD) decoder/driver is tied high, so this display is automatically blanked in the presence of a zero, under which condition the RBO terminal is high.

Consequently, the RBI terminal of the hundreds IC is high, so its display reads 2, and the RBO terminal is low. The RBI input of the tens unit is thus also low, so its display reads 0 and its RBO output is low. The least significant digit (LSD) is that of the units readout, and this does not require zero suppression; consequently, its RBI input is grounded and it reads 7. The display thus gives a total reading of 207.

Note in the *Figure 4.16* leading zero suppression circuit that ripple blanking feedback is applied backwards, from the MSD to the LSD. *Figure 4.17* shows how trailing zero suppression can be obtained by reversing the direction of feedback, from the LSD to the MSD. Thus, when an input of 1.1 V is fed to this circuit the LSD is blanked, since its BCD input is 0000 and its RBI input is high. Its RBO terminal is high under this condition, so the 100ths digit is also blanked in the presence of a 0000 BCD input.

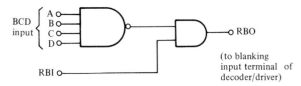

Figure 4.18   *DIY ripple-blanking logic (active-high type)*

Practical decoder/driver ICs are often (but not always) provided with ripple blanking input and output terminals; often, these are active low. If a decoder/driver IC does not incorporate integral ripple blanking logic, it can usually be obtained by adding external logic similar to that shown in *Figure 4.18*, with the RBO terminal connected to the *blanking* input pin of the decoder/driver IC. In *Figure 4.18* (an active high circuit), the output of the four-input NOR gate goes high only in the presence of a 0000 BCD input, and the RBO output goes high only if the decimal zero input is present while RBI is high.

Practical decoder/driver ICs

Practical decoder/driver ICs are available in both TTL and CMOS forms. Some of these devices have integral ripple blanking facilities, others have built-in data latches, and a few even have built-in decade counter stages, etc. Let us look at a few of the most popular of these devices.

The 7447A and 7448
These seven-segment decoder/driver ICs are members of the standard TTL family. They are also available in LS form under the designations 74LS47 and

74LS48 respectively. All of these ICs have integral ripple-blanking facilities, but do not incorporate data latches. *Figure 4.19* shows the outline and pin designations that are common to these devices, each of which is housed in a 16-pin DIL package.

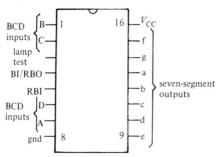

Figure 4.19 *Outline and pin designations of the 7447A, 74LS47, and 74LS48 range of BCD-to-seven-segment decoder/driver ICs*

The 7447A/74LS47 has an active-low output designed for driving a common-anode LED display via external current-limiting resistors (R_x), as shown in *Figure 4.20*. The 7448/74LS48 has an active-high output designed for driving a common-cathode LED display in a manner similar to that of *Figure 4.20*, but with the common terminal of the display taken to ground. In

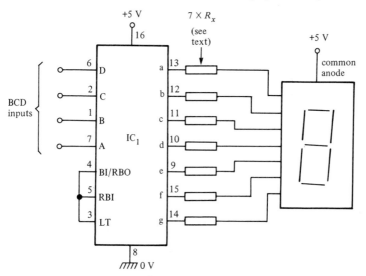

Figure 4.20 *Basic method of using the 7447A or 74LS47 to drive a common-anode LED display*

all cases, the R_x current-limiting resistors should be chosen to limit the segment currents below the following absolute limits:

$$7447A = 40 \text{ ma}; \quad 74LS47 = 24 \text{ mA}; \quad 7448/74LS48 = 6 \text{ mA}.$$

Figure 4.21 shows how the 7448/74LS48 can be used to drive a liquid-crystal display (LCD), using a pair of 7486 or 74LS86 quad two-input EX–OR gate ICs and an external 50 Hz square wave to apply the necessary phase signals to the display, as described earlier.

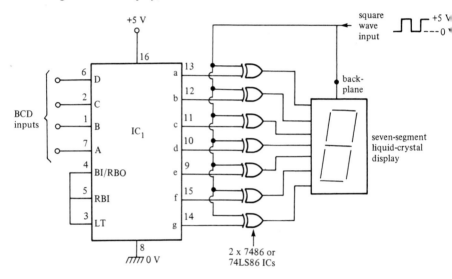

Figure 4.21   *Basic method of using the 7448 or 74LS48 to drive a liquid-crystal display*

Note from *Figure 4.19* that each of these ICs has three input 'control' terminals, these being designated LAMP TEST, BI/RBO, and RBI. The LAMP TEST terminal drives all display terminals on when the terminal is driven to logic low with the RBO terminal open or at logic high. When the BI/RBO terminal is pulled low all outputs are blanked; this pin also functions as a ripple-blanking output terminal. *Figure 4.22* shows how to connect the ripple-blanking terminals to give leading zero suppression on the first three digits of a four-digit display.

The 4511B

This BCD-to-seven-segment decoder IC has an integral four-bit data latch, but has no built-in facility for ripple-blanking. It is constructed with CMOS logic but features *npn* bipolar output transistor stages that are capable of sourcing output currents of up to 25 mA. *Figure 4.23* shows the outline and

BCD inputs

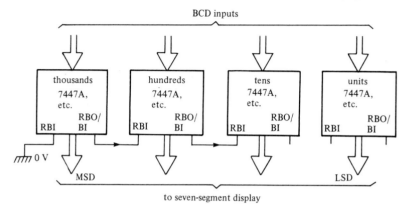

Figure 4.22 *Method of applying leading-zero suppression to the first three digits of a four-digit display, using the 7447A family of ICs*

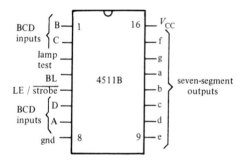

Figure 4.23 *Outline and pin designations of the 4511B CMOS BCD-to-seven-segment decoder/driver IC with integral four-bit data latch*

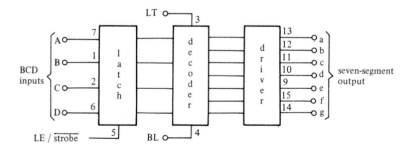

Figure 4.24 *Functional diagram of the 4511B IC*

pin notations of the device, and *Figure 4.24* shows the functional diagram of the IC, which can be used with any power source in the 5 V to 18 V range.

The 4511B has three control input terminals, these being designated LT (LAMP TEST), BL (BLANKING), and LE (LATCH ENABLE)/STROBE. The LT and BL inputs are active-low, and the LE input is active-high. In normal operation, LT and BL are taken high and LE is held low.

When the LE terminal is low, BCD input signals are decoded and fed directly to the seven-segment output terminals. If LE goes high the BCD input signals that are present at the moment of transition are latched into memory and fed (in decoded form) to the seven-segment outputs while LE remains high.

If the LT input is grounded, all output segments are activated, irrespective of the BCD inputs. If the BL input is grounded (with LT positive), all output segments are blanked.

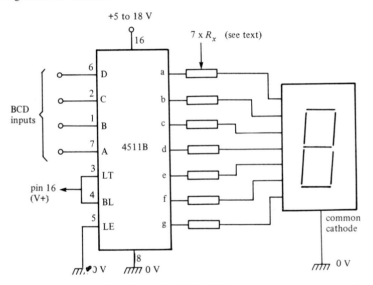

Figure 4.25   *Basic method of using the 4511B to drive a seven-segment common-cathode LED display*

The 4511B can be used to drive most popular types of seven-segment display. *Figure 4.25* shows the basic connections for driving a common-cathode LED display; a current-limiting resistor (R_x) must be wired in series with each display segment, and must have its value chosen to limit the segment current below 25 mA. Note that the 'segment' outputs of the 5411B are not internally current-limited, and the device thus has no output overload protection.

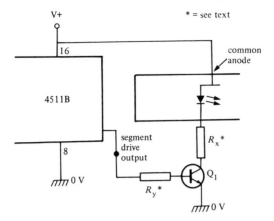

Figure 4.26 *Driving a common-anode LED display*

Figures 4.26, 4.27 and *4.28* show how to modify the above circuit to drive LED common-anode displays, gas discharge displays, and low-brightness fluorescent displays respectively. Note in the cases of *Figures 4.26* and *4.27* that an *npn* buffer transistor must be interposed between each output drive segment and the input segment of the display; in each case, R_x determines the operating segment current of the display, and R_y determines the base current of the transistor.

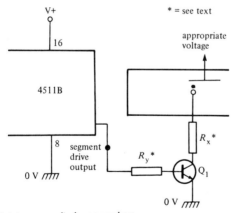

Figure 4.27 *Driving a gas discharge readout*

The 4511B can also be used to drive seven-segment liquid-crystal displays by using an external square-wave phase signal and a set of EX-OR gates in a configuration similar to that of *Figure 4.21*. In practice, however, it is far better to use a 4543B IC for this particular application.

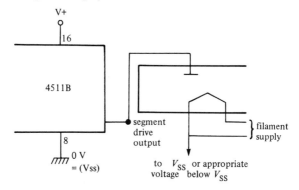

Figure 4.28   *Driving a low-brightness fluorescent readout*

The 4543B
This seven-segment decoder/driver IC is a CMOS device with integral four-bit data latch. It is specifically designed for driving liquid-crystal displays, but can also drive most other types of seven-segment display. *Figure 4.29* shows the outline and pin designations of the device, which can be used with any power supply in the 3 V to 18 V range.

The 4543B has three input control terminals, these being designated LD (LATCH DISABLE), PHASE, and BL (BLANK). In normal use the LD terminal is biased high and the BL terminal is tied low. The state of the PHASE terminal depends on the type of display that is being driven. For

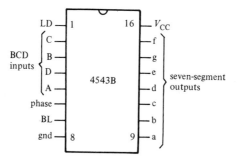

Figure 4.29   *Outline and pin designations of the 4543B BCD-to-seven-segment latch/decoder/driver for liquid crystals*

driving LCD readouts, a square wave (roughly 50 Hz, swinging fully between the GND and V_{cc} values) must be applied to the PHASE terminal: for driving common-cathode LED displays, PHASE must be grounded: for driving common-anode displays, PHASE must be tied to logic high.

The display can be blanked at any time by simply driving the BL terminal to the logic-high state. When the LD terminal is in its normal high state, BCD inputs are decoded and fed directly to the seven-segment output terminals of the IC. When the LD terminal is pulled low, the BCD input signals that are present at the moment of transition are latched into memory and fed (in decoded form) to the seven-segment outputs while LD remains low.

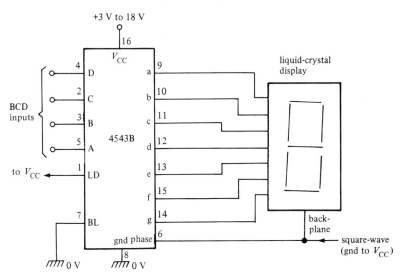

Figure 4.30 *Basic method of using the 4543B to drive an LCD*

Figure 4.30 shows the basic method of using the 4543B to drive an LCD, and *Figures 4.31* to *4.34* show how this ciruit can be modified to drive other types of seven-segment display. Note in the cases of *Figures 4.31* and *4.32* that the values of R_x must be chosen to limit the output drive currents to below 10 mA per segment; if higher drive currents are needed, interpose a buffer transistor between the output of the 4543B and the input of the display segment.

The 4026B

This device is a complete decade counter with integral decoder/driver circuitry that can directly drive a seven-segment common-cathode LED display. The segment output currents are internally limited (to about 5 mA at 10 V supply, 10 mA at 15 V supply), so the display can be connected directly to the output of the IC without the use of external current-limiting resistors. The device does not incorporate a data latch and has no facility for ripple blanking. *Figure 4.35* shows its outline and pin designations.

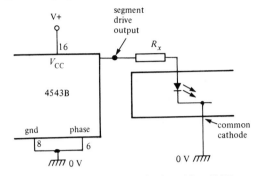

Figure 4.31   *Driving a common-cathode LED display with a 4543B*

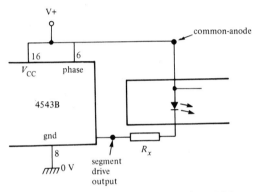

Figure 4.32   *Driving a common-anode LED display with a 4543B*

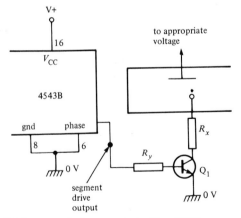

Figure 4.33   *Driving a gas-discharge readout with a 4543B*

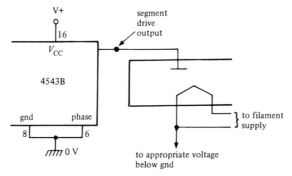

Figure 4.34 *Driving a fluorescent readout with a 4543B*

The 4026B has four input control terminals, and three auxiliary output terminals. The input terminals are designated CLK (CLOCK), CI (CLOCK INHIBIT), RESET, and D/E (DISPLAY ENABLE). The IC incorporates a Schmitt trigger on its CLOCK input line, and clock signals do not have to be pre-shaped. The counter is reset to zero by driving the RESET terminal high.

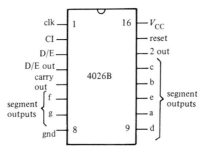

Figure 4.35 *Outline and pin designations of the 4026B decade counter with seven-segment outputs*

The CI terminal must be grounded to allow normal counting operation: when CI is high the counters are inhibited. The display is blanked when the D/E terminal is grounded: the D/E terminal must be high for normal operation. Thus, in normal operation the RESET and CI terminals are grounded and the D/E terminal is held positive, as shown in *Figure 4.36.*

The three auxiliary output terminals of the 4026B are designated D/E OUT, CARRY OUT, and 2 OUT. The D/E OUT signal is a slightly delayed copy of the D/E input signal. The CARRY OUT signal is a symmetrical square wave at one tenth of the CLOCK input frequency, and is useful in cascading 4026B counters. The 2 OUT terminal goes low only on a count of 2. *Figure 4.36* shows the basic circuit connections to be used when cascading stages.

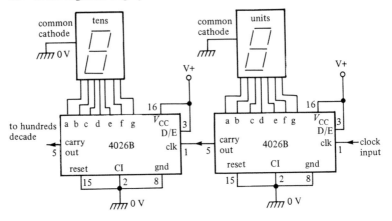

Figure 4.36   *Basic method of cascading 4026B ICs*

The 4033B

This device (see *Figure 4.37*) can be regarded as a modified version of the 4026B, with the D/E and D/E OUT terminals eliminated and replaced by ripple blanking input (RBI) and output (RBO) terminals, and with the 2 OUT terminal replaced with a LT (LAMP TEST) terminal which activates all output segments when biased high. In normal use the RESET, CI and LT

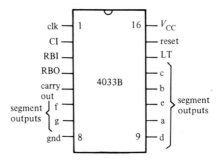

Figure 4.37   *Outline and pin designations of the 4033B decade counter with seven-segment outputs and ripple blanking*

terminals are all grounded and the RBI terminal is made positive, as shown in *Figure 4.38*: this configuration does not provide blanking of unwanted leading and/or trailing zeros.

If cascaded 4033B ICs are required to give automatic leading-zero suppression the basic *Figure 4.38* circuit must be modified as shown in *Figure 4.39*, to provide ripple-blanking operation. Here, the RBI terminal of the

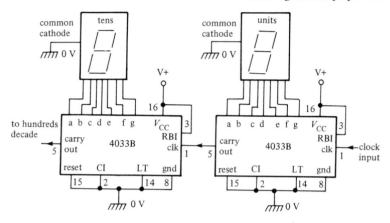

Figure 4.38   *Basic method of cascading 4033B ICs (without zero suppression)*

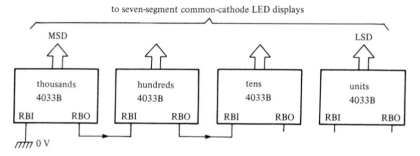

Figure 4.39   *Method of modifying the Figure 4.38 circuit to give automatic leading-zero suppression*

most significant digit (MSD) is grounded, and its RBO terminal is connected to the RBI terminal of the next least-significant stage. This procedure is repeated on all except the LSD, which does not require zero suppression. If trailing-zero suppression is required, the direction of ripple-blanking feedback must be reversed, with the RBI terminal of the LSD grounded and its RBO terminal wired to the RBI terminal of the next least-significant stage, and so on.

5   Light-sensitive devices

In Chapters 2, 3 and 4 we have dealt with the theory and applications of light-generating and light-reflecting devices, such as LEDs and LCDs. In this chapter we concentrate on the operating principles and applications of a variety of light sensitive devices, such as LDRs and photodiodes, etc.

LDR basics

Electronic optosensors are devices that alter their electrical characteristics, in the presence of visible or invisible light. The best known devices of these types are the LDR (light dependent resistor), the photodiode, and the phototransistor. Let us start off by concentrating on the LDR and LDR circuitry.

LDR operation relies on the fact that the conductive resistance of a film of cadmium sulphide (CdS) varies with the intensity of light falling on the face of the film. This resistance is very high under dark conditions and low under bright conditions. *Figure 5.1* illustrates the basic construction of the LDR,

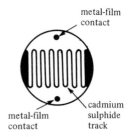

Figure 5.1   *Basic structure of the LDR*

and *Figure 5.2* shows the LDR symbol. The device consists of a pair of metal film contacts separated by a snake-like track of cadmium sulphide film, designed to provide the maximum possible contact area with the two metal films. The structure is housed in a clear plastic or resin case, to provide free access to external light.

Figure 5.2   *LDR symbol*

Practical LDRs are available in a variety of sizes and package styles, the most popular size having a face diameter of roughly 10 mm. *Figure 5.3* shows the typical characteristic curve of such a device, which has a resistance of about 900R at a light intensity of 100 lx (typical of a well lit room) or about 30R at an intensity of 8000 lx (typical of bright sunlight). The resistance rises to several megohms under dark conditions.

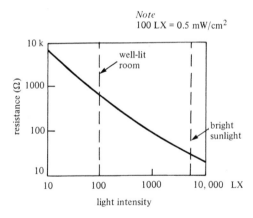

Figure 5.3   *Typical characteristics curve of a LDR with a 10 mm face diameter*

LDRs are sensitive, inexpensive, and readily available devices. They have good power and voltage handling capabilities, similar to those of a conventional resistor. Their only significant defect is that they are fairly slow acting, taking tens or hundreds of milliseconds to respond to sudden changes in light level. Useful practical LDR applications include light- and dark-activated switches and alarms, light-beam alarms, and reflective smoke alarms, etc. *Figures 5.4* to *5.19* show some practical applications of the device; each of these circuits will work with virtually any LDR with a face diameter in the range 3 mm to 12 mm.

LDR light-switches

Figures 5.4 to *5.9* show some practical relay-output light-activated switch circuits based on the LDR. *Figure 5.4* shows a simple non-latching circuit, designed to activate when light enters a normally dark area such as the inside of a safe or cabinet, etc.

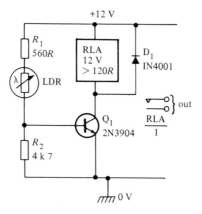

Figure 5.4   *Simple non-latching light-activated relay switch*

Here, R_1-LDR and R_2 form a potential divider that controls the base-bias of Q_1. Under dark conditions the LDR has a very high resistance, so zero base-bias is applied to Q_1, and Q_1 and RLA are off. When a significant amount of light falls on the LDR face the LDR resistance falls to a fairly low value and base-bias is applied to Q_1, which thus turns on and activates the RLA/1 relay contacts, which can be used to control external circuitry.

The simple *Figure 5.4* circuit has a fairly low sensitivity, and has no facility for sensitivity adjustment. *Figure 5.5* shows how these defects can be overcome by using a Darlington-connected pair of transistors in place of Q_1, and by using sensitivity control RV_1 in place of R_2. The diagram also shows how the circuit can be made to give a self-latching action via relay contacts RLA/2; normally-closed push-button switch SW_1 enables the circuit to be reset (unlatched) when required.

Figure 5.6 shows how a LDR can be used to make a simple dark-activated relay switch that turns on when the light level falls below a value pre-set via

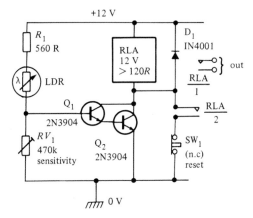

Figure 5.5   *Sensitive self-latching light-activated relay switch*

RV_1. Here, R_1 and LDR form a potential divider that generates an output voltage that rises as the light level falls. This voltage is buffered by emitter-follower Q_1 and used to control relay RLA via common-emitter amplifier Q_2 and current-limiting resistor R_3.

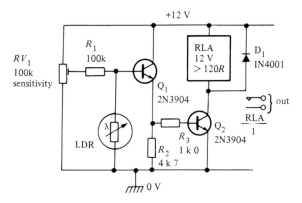

Figure 5.6   *Simple dark-activated relay switch*

The light trigger points of the *Figure 5.4* to *5.6* circuits are susceptible to variations in circuit supply voltage and ambient temperature. *Figure 5.7* shows a very sensitive precision light-activated circuit that is not influenced by such variations. In this case, LDR–RV_1 and R_1–R_2 are connected in the form of a Wheatstone bridge, and the op-amp and Q_1–RLA act as a highly sensitive balance-detecting switch. The bridge balance point is quite independent of variations in supply voltage and temperature, and is influenced only by variations in the relative values of the bridge components.

In *Figure 5.7*, the LDR and RV_1 form one arm of the bridge, and R_1–R_2 form the other arm. These arms can actually be regarded as potential dividers, with the R_1–R_2 arm applying a fixed half-supply voltage to the non-inverting input of the op-amp, and with the LDR–RV_1 divider applying a light-dependent variable voltage to the inverting terminal of the op-amp.

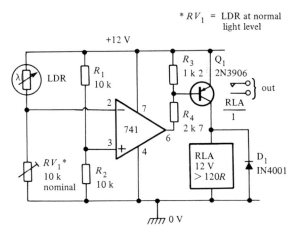

$* RV_1$ = LDR at normal light level

Figure 5.7   *Precision light-sensitive relay switch*

In use, RV_1 is adjusted so that the LDR–RV_1 voltage rises fractionally above that of R_1–R_2 as the light intensity rises to the desired trigger level, and under this condition the op-amp output switches to negative saturation and thus drives the relay on via Q_1 and biasing resistors R_3–R_4. When the light intensity falls below this level, the op-amp output switches to positive saturation, and under this condition Q_1 and the relay are off.

The *Figure 5.7* circuit is very sensitive, being able to detect light-level changes too small to be seen by the human eye. The circuit can be modified to act as a precision dark-activated switch by either transposing the inverting and non-inverting input terminals of the op-amp, or by transposing RV_1 and the LDR. *Figure 5.8* shows a circuit using the latter option.

Figure 5.8 also shows how a small amount of hysteresis can be added to the circuit via feedback resistor R_5, so that the relay turns on when the light level falls to a particular value, but does not turn off again until the light intensity rises a substantial amount above this value. The magnitude of hysteresis is inversely proportional to the R_5 value, being zero when R_5 is open circuit.

A precision combined light/dark switch, which activates a single relay if the light intensity rises above one pre-set value or falls below another pre-set

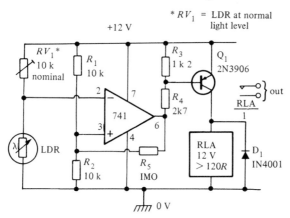

Figure 5.8   *Precision dark-activated switch, with hysteresis*

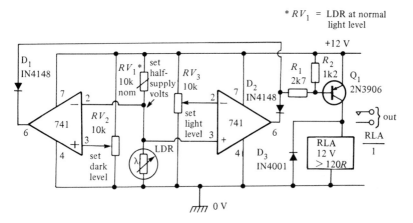

Figure 5.9   *Combined light-/dark-activated switch with single relay output*

value, can easily be made by combining an op-amp light switch and an op-amp dark switch in the manner shown in *Figure 5.9*.

To set up the *Figure 5.9* circuit, first pre-set RV_1 so that approximately half-supply volts appear on the LDR–RV_1 junction when the LDR is illuminated at the mean or normal intensity level. RV_2 can then be pre-set so that RLA turns on when the light intensity falls to the desired dark level, and RV_3 can be adjusted so that RLA activates at the desired brightness level.

Note in the *Figure 5.7* to *5.9* circuits that the adjusted RV_1 value should equal the LDR resistance value at the normal light level of each circuit.

Bell-output LDR alarms

The *Figure 5.4* to *5.9* light-activated LDR circuits all have relay outputs which can be used to control virtually any type of external circuitry. In many light-activated applications, however, circuits are required to act as audible-output alarms, and this type of action can be obtained without the use of relays. *Figures 5.10* to *5.16* show some practical circuits of this type.

The *Figure 5.10* to *5.12* circuits are each designed to give a direct output to an alarm bell (or buzzer). This bell must be of the self-interrupting type and must consume an operating current of less than 2 A. The supply voltage of each circuit should be 1.5 to 2 V greater than the nominal operating value of the bell.

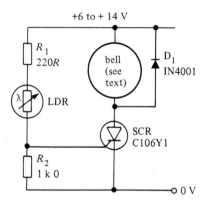

Figure 5.10   *Simple light-activated alarm bell*

Figure 5.10 shows the circuit of a simple light-activated alarm; the operating theory is fairly simple. The LDR and R_2 form a potential divider: under dark conditions the LDR resistance is high, so the LDR–R_2 junction voltage is too small to activate the gate of the SCR (silicon controlled rectifier), but under bright conditions the LDR resistance is low, so gate bias is applied to the SCR, which turns on and activates the alarm bell.

Note in the above circuit that, although the SCR is a self-latching device, the fact that the bell is of the self-interrupting type ensures that the SCR automatically unlatches repeatedly as the bell operates (and the SCR anode current falls to zero in each self-interrupt phase). Consequently, the alarm bell automatically turns off again when the light level falls back below the trip level.

The *Figure 5.10* circuit has a fairly low sensitivity and has no facility for sensitivity adjustment. *Figure 5.11* shows how these defects can be overcome by using RV_1 in place of R_2 and by using Q_1 as a buffer between the LDR and the SCR gate. This diagram also shows how the circuit can be made self-latching by wiring R_4 across the bell so that the SCR anode current does not fall to zero as the bell self-interrupts. Switch SW_1 enables the circuit to be reset (unlatched) when required.

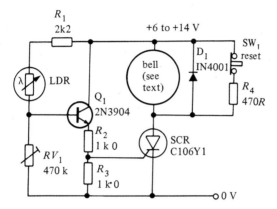

Figure 5.11   *Improved light-activated alarm bell with self-latching facility*

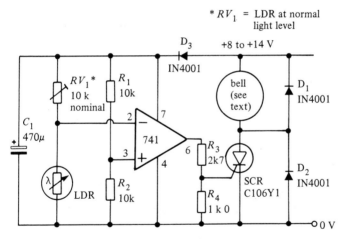

Figure 5.12   *Precision light-activated alarm bell*

Figure 5.12 shows how to make a precision light-alarm with a SCR-driven alarm bell output, by using a Wheatstone bridge ($LDR-RV_1-R_1-R_2$) and an op-amp balance detector to give the precision action. This circuit can be converted into a dark-activated alarm by simply transposing RV_1 and the LDR. Hysteresis can also be added, if required.

98 Light-sensitive devices

Speaker-output LDR alarms

Figures 5.13 to *5.16* show various ways of using CMOS 4001B quad two-input NOR gate ICs to make light-activated alarms that generate audible outputs in loudspeakers. The *Figure 5.13* circuit is that of a dark-activated alarm that generates a low-power 800 Hz pulsed-tone signal in the speaker. Here, IC_{1c} and IC_{1d} are wired as a 800 Hz astable multivibrator that can feed tone signals into the speaker via Q_1 and is gated on only when the output of IC_{1b} is low, and IC_{1a}–IC_{1b} are wired as a 6 Hz astable that is gated on only when its pin 1 gate terminal (which is coupled to the LDR–RV_1 potential divider) is pulled low.

The action of the *Figure 5.13* circuit is as follows. Under bright conditions the LDR–RV_1 junction voltage is high, so both astables are disabled and no signal is generated in the speaker. Under dark conditions the LDR–RV_1 junction voltage is low, so the 6 Hz astable is activated and in turn gates the 800 Hz astable on and off at a 6 Hz rate, thereby generating a pulsed-tone signal in the speaker via Q_1.

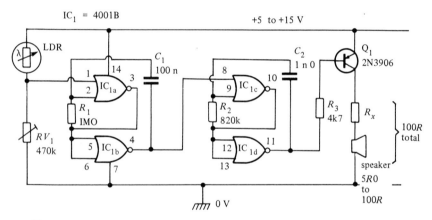

Figure 5.13 *Dark-activated alarm with pulsed-tone output*

The precise switching or gate point of the 4001B IC is determined by the threshold voltage value of the IC, and this is a percentage value of the supply voltage: the value is nominally 50%, but may vary from 30% to 70% between individual ICs. In practice, the switching point of each individual 4001B IC is very stable, and the *Figure 5.13* circuit gives very sensitive dark-activated alarm triggering.

Figure 5.14 shows the circuit of a self-latching light-activated alarm with a 800 Hz monotone output. In this case IC_{1c}–IC_{1d} are again wired as a gated 800 Hz astable, but IC_{1a}–IC_{1b} are wired as a bistable multivibrator with an output that (under dark conditions) is normally high, thus gating the 800 Hz astable off. Under bright conditions, however, the LDR–RV_1 junction goes high and latches the bistable into its alternative 'output low' state, thereby gating the 800 Hz astable on and generating the monotone alarm signal; once latched, the circuit remains in this on state until dark conditions return and the bistable is simultaneously reset via SW_1.

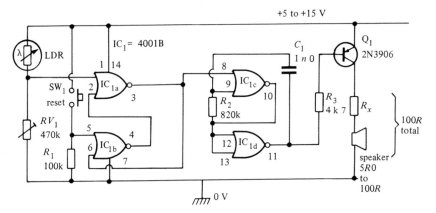

Figure 5.14 *Self-latching light-activated alarm with monotone output*

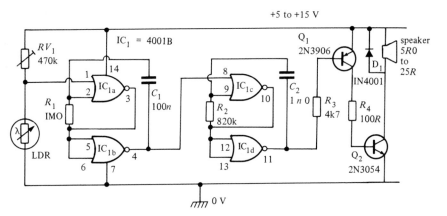

Figure 5.15 *Boosted-output pulsed-tone light-activated alarm*

Note that the light/dark operation of the *Figure 5.13* and *5.14* circuits can be reversed by simply transposing the LDR–RV_1 positions, and that each circuit gives an output power of only a few milliwatts. *Figure 5.15* shows how the operation of the *Figure 5.13* circuit can be reversed (to give light-operated

action) by transposing LDR and RV_1, and how the output power can be
boosted via an additional transistor. The circuit can be used with supply
voltages in the range 5–15 V, and with speaker impedances in the range $5R0$
to $25R$. The circuit gives output powers in the range 0.25 W to 11.25 W,
depending on the values of impedance and voltage that are used.

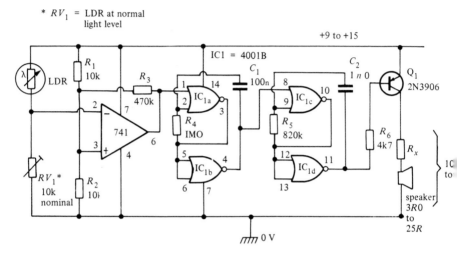

Figure 5.16 *Precision light-activated pulsed-tone alarm with hysteresis*

The *Figure 5.13* to *5.15* circuits have sensitivity levels that are adequate for
most practical purposes. If required, sensitivity can be further boosted, and
trigger-level stability can be increased, by interposing an op-amp voltage
comparator between the LDR–RV_1 light-sensitive potential divider and the
gate terminal of the CMOS waveform generator, as shown in *Figure 5.16*.
Resistor R_3 controls the hysteresis of the circuit, and can be removed if the
hysteresis is not needed.

Miscellaneous LDR circuits

Two of the best known applications of the LDR are as light-beam alarms and
switches, which activate when the passage of a beam of light is interrupted,
and as smoke alarms, which activate when smoke causes a light source to
reflect on to the face of an LDR. *Figures 5.17* to *5.19* show self-interrupting
alarm-bell versions of these circuits.

The light-beam alarm circuit of *Figure 5.17* acts like a dark-operated alarm.
Normally, the LDR face is illuminated by the light beam, so the LDR has a

low resistance and little voltage appears on the RV_1–LDR junction, and the SCR and bell are off. When the light beam is broken the LDR resistance increased, so a significant voltage appears on the RV_1–LDR junction and the SCR activates and turns on the alarm bell; R_3 is used to self-latch the alarm.

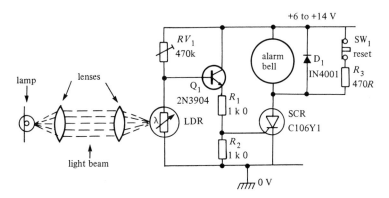

Figure 5.17   *Simple light-beam alarm with self-interrupting bell output*

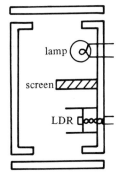

Figure 5.18   *Sectional view of reflection-type smoke detector*

Figure 5.18 shows a sectional view of a reflective-type smoke detector. Here, the lamp and LDR are mounted in an open ended but light-excluding box, in which an internal screen prevents the lamp-light from falling directly on the LDR face. The lamp is a source of both light and heat, and the heat causes convection currents of air to be drawn in from the bottom of the box and to be expelled through the top. The inside of the box is painted matt black, and the construction lets air pass through the box but excludes external light.

Thus, if the convected air currents are smoke free, no light falls on the LDR face, and the LDR presents a high resistance. If the air currents do contain

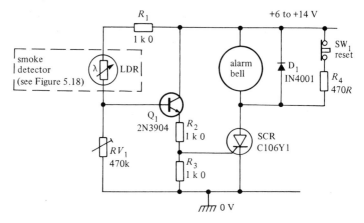

Figure 5.19 *Smoke alarm with self-interrupting bell output*

smoke, however, the smoke particles cause the light of the lamp to reflect on to the LDR face and so cause a great and easily detectable decrease in the LDR resistance. *Figure 5.19* shows the practical circuit of a reflection-type smoke alarm that can be used with this detector; the circuit acts like a light-operated alarm.

Photo-diodes

If a conventional silicon diode is connected in the reverse-biased circuit of *Figure 5.20*, negligible current will flow through the diode and zero voltage will develop across R_1. If the diode casing is now carefully removed so that the diode's semiconductor junction is revealed, and the junction is then exposed to visible light in the same circuit, the diode current will rise, possibly to as

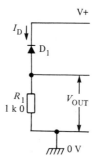

Figure 5.20 *Reverse-biased diode circuit*

high as 1 mA, producing a significant output across R_1. Further investigation will show that the diode current (and thus the output voltage) is directly proportional to light intensity, and that the diode is therefore photosensitive.

In practice, *all* silicon junctions are photosensitive, and a photodiode can be regarded as a conventional diode housed in a case that lets external light reach its photosensitive semiconductor junction. *Figure 5.21* shows the standard photodiode symbol. In use, the photodiode is reverse biased and the output voltage is taken from across a series-connected load resistor. This resistor may be connected between the diode and ground, as in *Figure 5.20*, or between the diode and the positive supply line, as in *Figure 5.22*.

Figure 5.21   *Photodiode symbol*

The human eye is sensitive to a range of light radiation, as shown in *Figure 5.23*. It has a peak spectral response to the colour green, which has a wave length of about 550 nm, but has a relatively low sensitivity to the colour violet (400 nm) at one end of the spectrum and to dark red (700 nm) at the other. Photodiodes also have spectral response characteristics, and these are determined by the chemistry used in the semiconductor junction material. *Figure 5.23* shows typical response curves of a general-purpose photodiode, and an infra-red (IR) photodiode.

Photodiodes have a far lower light-sensitivity than cadmium-sulphide LDRs, but give a far quicker response to changes in light level. Generally, LDRs are ideal for use in slow-acting direct-coupled light-level sensing

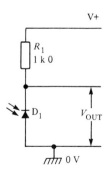

Figure 5.22   *Photodiode circuit with D_1-to-$V+$ load*

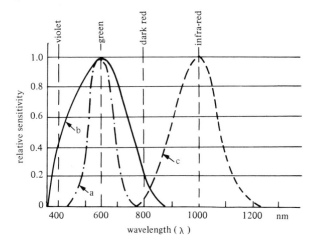

Figure 5.23   *Typical spectral response curves of (a) the human eye, (b) a general-purpose photodiode, and (c) an infra-red photodiode*

applications, while photodiodes are ideal for use in fast-acting AC-coupled signalling applications. Typical photodiode applications include IR remote-control circuits, IR beam switches and alarm circuits, and photographic flash slave circuits, etc.

Phototransistors

Figure 5.24 shows the standard symbol of a phototransistor, which can be regarded as a conventional transistor housed in a case that enables its semiconductor junctions to be exposed to external light. The device is normally used with its base open circuit, in either of the configurations shown in *Figure 5.25*, and functions as follows.

Figure 5.24   *Phototransistor symbol*

In *Figure 5.25(a)*, the base-collector junction of the transistor is effectively reverse biased and thus acts as a photodiode. The photo-generated currents of the base-collector junction feed directly into the base of the device, and the normal current-amplifying transistor action causes the output current to appear (in greatly amplified form) as collector current, and in *Figure 5.25(a)*, R_1 causes this current to generate an output voltage as shown.

In practice, the collector and emitter currents of the transistor are virtually identical and, since the base is open circuit, the device is not subjected to significant negative feedback. Consequently, the alternative *Figure 5.25(b)* circuit, in which R_1 is connected to Q_1 emitter, gives a virtually identical performance to that of *Figure 5.25(a)*.

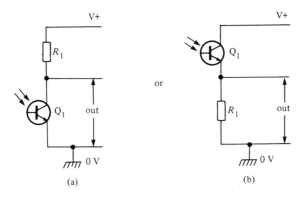

Figure 5.25   *Alternative phototransistor configurations*

The sensitivity of a phototransistor is typically one hundred times greater than that of a photodiode, but its useful maximum operating frequency (a few hundred kilohertz) is proportionally lower than that of a photodiode (tens of megahertz). A phototransistor can be converted into a photodiode by using only its base and collector terminals and ignoring the emitter, as shown in *Figure 5.26*.

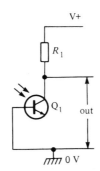

Figure 5.26   *Phototransistor used as a photodiode*

Alternatively, the sensitivity (and operating speed) of a phototransistor can be made variable by wiring a variable resistor between the base and emitter, as shown in *Figure 5.27*. With RV_1 open circuit, phototransistor operation is obtained; with RV_1 short circuit, photodiode operation occurs.

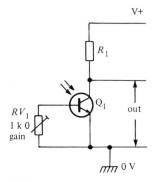

Figure 5.27   *Variable-sensitivity phototransistor circuit*

Note in the *Figure 5.20* to *5.27* circuits that, in practice, the R_1 load value is usually chosen on a compromise basis, since the circuit voltage gain increases but the useful operating bandwidth decreases as the R_1 value is increased. Also, the R_1 value must, in many applications, be chosen to bring the photo-sensitive device into its linear operating region.

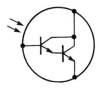

Figure 5.28   *Darlington phototransistor symbol*

Some phototransistors are constructed in Darlington or Super-Alpha form, and use the symbol shown in *Figure 5.28*. These devices have typical sensitivities some ten times greater than a normal phototransistor, but have useful maximum operating frequencies of only tens of kilohertz. Devices are also available with integral amplifiers, and in the form of photosensitive SCRs and triacs, etc.

Pre-amp circuits

Photodiodes and phototransistors are often used as the sensing elements at the receiver end of an opto data transfer system, such as a light-beam switch or alarm or remote control system, etc., in which data is sent to the receiver via an opto carrier wave.

In such applications, the signal reaching the photosensor may at some times be very weak, and at other times very strong. Also, the sensor may be subjected to a great deal of noise in the form of unwanted light (visible or invisible) signals, etc. To help minimize these problems, the link is usually operated in the infra-red range, and the optosensor output is passed to processing circuitry via a low-noise pre-amplifier with a wide dynamic operating range. *Figures 5.29* and *5.30* show typical examples of such circuits, using photodiode sensors.

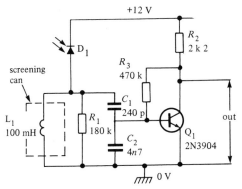

Figure 5.29   *Selective IR pre-amp designed for 30 kHz operation*

The *Figure 5.29* circuit is designed for use with a 30 kHz carrier wave, and tuned circuit $L_1-C_1-C_2$ is wired in series with D_1 and damped by R_1 to provide the necessary frequency-selective low-noise action. The output signals are tapped off at the C_1-C_2 junction and then amplified by Q_1.

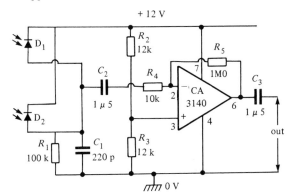

Figure 5.30   *20 kHz selective pre-amp for use in beam alarm applications*

Finally, to complete this chapter, *Figure 5.30* shows a 20 kHz selective pre-amplifier circuit for use in an IR light-beam alarm application, in which the alarm sounds when the beam is broken. Here, two IR photodiodes are wired in parallel, so that beam signals are lost only when *both* diode signals are cut off, and share a common 100 k load resistor (R_1). This resistor is shunted by C_1 to reject unwanted high-frequency signals, and the R_1 output signals are fed to the × 100 op-amp inverting amplifier via C_2, which rejects unwanted low-frequency signals.

6 Optocoupler devices

In Chapters 2, 3, 4 and 5 we have dealt with light-generating devices such as
the LED, and with light-sensitive devices such as the phototransistor. In this
chapter we deal with the so-called optocoupler range of devices, which
incorporate both an LED and a phototransistor in a single package and have
a wide range of practical applications.

Optocoupler basics

An LED is a light-generating device, and a phototransistor is a light-sensitive
device. Consequently, if the two devices are mounted close together in a single
light-excluding package so that the LED light can fall on the phototransistor
face, as shown in *Figure 6.1*, and the device is then connected into the circuit of
Figure 6.2, it will be found that the conduction current of Q_1 can be controlled
via the conduction current of the LED, even though the two devices are
physically separated. Such a package is known as an optocoupler, since the
input (the LED) and the output (the phototransistor) devices are optically
coupled.

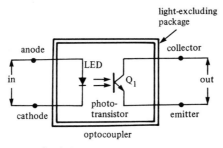

Figure 6.1   *Basic optocoupler device*

Thus, in *Figure 6.2*, when SW$_1$ is open no current flows in the LED, so no light falls on the face of Q$_1$, so Q$_1$ is virtually open-circuit and zero voltage is developed across output resistor R_2. When SW$_1$ is closed, current flows through the LED via R_1, and the resulting light falls on Q$_1$ face, causing the phototransistor to conduct and generate an output voltage across R_2.

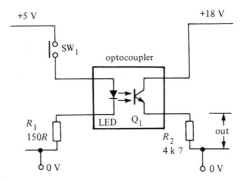

Figure 6.2   *Basic optocoupler circuit*

Note that the simple optically-coupled circuit of *Figure 6.2* can be used with digital input/output signals only, but that in practice the circuit can easily be modified for use with analogue input/output signals; we show how later.

The most important point to note about the optocoupler device of *Figure 6.1* is that a circuit connected to its input can be electrically fully isolated from the output circuit, and that a potential difference of hundreds or thousands of volts can safely exist between these two circuits without adversely influencing the optocoupler action. This isolating characteristic is the main attraction of this type of optocoupler device, which is generally known as an isolating optocoupler.

Typical isolating optocoupler applications include low-voltage to high-voltage (or vice versa) signal coupling; interfacing of a computer output signal to external electronic circuitry or electric motors, etc.; interfacing of ground-referenced low-voltage circuitry to floating high-voltage circuitry driven directly from the mains AC power lines, etc. Optocouplers can also be used to replace low-power relays and pulse transformers in many applications.

Special optocouplers

The *Figure 6.1* device is a simple isolating optocoupler. *Figures 6.3* and *6.4* show two other types of optocoupler. The device shown in *Figure 6.3* is known as a slotted optocoupler, and has a slot moulded into the package between the LED light source and the phototransistor light sensor.

Here, light can normally pass from the LED to Q_1 without significant attenuation by the slot. The optocoupling can, however, be completely blocked by placing an opaque object in the slot. The slotted optocoupler can thus be used in a variety of presence detecting applications, including end-of-tape detection, limit switching, and liquid-level detection.

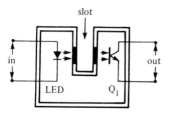

Figure 6.3   *Slotted optocoupler device*

The device shown in *Figure 6.4* is known as a reflective optocoupler. Here, the LED and Q_1 are optically screened from each other within the package, and both face outwards (in the same direction) from the package. The construction is such that an optocoupled link can be set up by a reflective

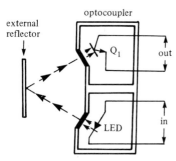

Figure 6.4   *Reflective optocoupler*

object (such as metallic paint or tape, or even smoke particles) placed a short distance outside the package, in line with both the LED and Q_1 The reflective optocoupler can thus be used in applications such as tape-position detection, engine-shaft revolution counting or speed measurement, or smoke or fog detection, etc.

Transfer ratios

One of the most important parameters of an optocoupler device is its optocoupling efficiency, and to maximize this parameter the LED and the phototransistor (which usually operate in the infra-red range) are always closely matched spectrally.

The most convenient way of specifying optocoupling efficiency is to quote the output-to-input current transfer ratio (CTR) of the device, i.e., the ratio of the output current (I_C) measured at the collector of the phototransistor, to the input current (I_F) flowing into the LED. Thus, CTR = I_C/I_F. In practice, CTR may be expressed as a simple figure such as 0.5, or (by multiplying this figure by 100) as a percentage figure such as 50%.

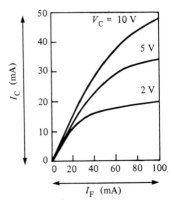

Figure 6.5   *Typical I_C/I_F characteristics of a simple optocoupler at various values of output-transistor collector voltage*

Simple isolating optocouplers with single-transistor output stages have typical CTR values in the range 20% to 100%; the actual CTR value depends (among other things) on the input and output current values of the device and on the supply voltage value of the optotransistor. *Figure 6.5* shows three typical sets of output/input currents obtained at different V_C values.

It should be noted that, because of variations in LED radiation efficiency and phototransistor current gains, the actual CTR values of individual optocouplers may vary significantly from the typical value. An optocoupler type with a typical CTR value of 60% may, for example, in fact have a true value in the range 20% to 180% in an individual device.

Other parameters

Other important optocoupler parameters include the following.

Isolation voltage
This is the maximum permissible DC potential that can be allowed to exist between the input and output circuits. Typical values vary from 500 V to 4 kV.

$V_{CE(MAX)}$
This is the maximum allowable DC voltage that can be applied across the output transistor. Typical values vary from 20 V to 80 V.

$I_{E(MAX)}$
This is the maximum permissible DC current that can be allowed to flow in the input LED. Typical values vary from 40 mA to 100 mA.

Bandwidth
This is the typical maximum signal frequency (in kilohertz) that can be usefully passed through the optocoupler when the device is operated in its normal mode. Typical values vary from 20 kHz to 500 kHz, depending on the type of device construction.

Practical optocouplers

Optocouplers are produced by several different manufacturers. They are available in a limited number of basic forms, but are retailed under a vast number of different type numbers. Rather than list all of these types individually, we will simply look here at typical examples of these devices.

Practical optocoupler devices are available in six basic forms, and these are illustrated in *Figures 6.6* to *6.11*. Four of these devices (*Figures 6.6* to *6.9*) are isolating optocouplers, and the remaining two are the slotted optocoupler (*Figure 6.10*) and the reflective optocoupler (*Figure 6.11*). The table in *Figure 6.12* lists the typical parameter values of these six devices.

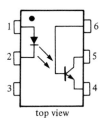

top view

Figure 6.6 *Typical simple isolating optocoupler*

The simple isolating optocoupler (*Figure 6.6*) uses a single phototransistor output stage and is usually housed in a six-pin package, with the base terminal of the phototransistor externally available. In normal use the base is left open circuit, and under this condition the optocoupler has a minimum CTR value of 20% and a useful bandwidth of 300 kHz. The phototransistor can, however, be converted to a photodiode by shorting the base (pin 6) and emitter (pin 4) terminals together; under this condition the CTR value falls to about 0.2% but the bandwidth rises to about 30 MHz.

The Darlington optocoupler (*Figure 6.7*) is also housed in a six-pin package and has its phototransistor base externally available. Because of the high current gain of the Darlington, this coupler has a typical minimum CTR value of about 300%, but has a useful bandwidth of only 30 kHz.

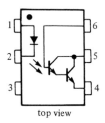

top view

Figure 6.7   *Typical Darlington isolating optocoupler*

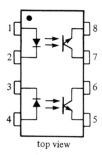

top view

Figure 6.8   *Typical dual isolating optocoupler*

The dual and quad optocouplers of *Figures 6.8* and *6.9* both use single-transistor output stages in which the base terminal is not externally available.

Note in all four isolating devices that the input pins are on one side of the package and the output pins are on the other. This construction facilitates the maximum possible values of isolating voltage. Also note in the multichannel devices of *Figures 6.8* and *6.9* that, although these devices have isolating voltage values of 1.5 kV, potentials greater than 500 V should not be allowed to exist between adjacent channels.

Isolating voltage values are not specified for the slotted and reflective optocoupler devices of *Figures 6.10* and *6.11*. The *Figure 6.10* device has a typical slot width of about 3 mm, and uses a single output transistor to give an open slot minimum CTR value of 10% and a bandwidth of 300 kHz.

Finally, the reflective optocoupler of *Figure 6.11* uses a Darlington output

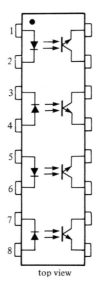

Figure 6.9   *Typical quad isolating optocoupler*

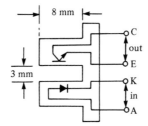

Figure 6.10   *Typical slotted optocoupler*

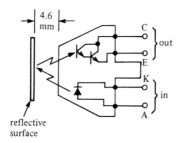

Figure 6.11   *Typical reflective optocoupler*

parameter	isolating optocouplers				slotted optocoupler	reflective opto-coupler
	simple type	Darlington type	dual type	quad type		
isolating voltage	± 4 kV	±4 kV	± 1.5 kV	± 1.5 kV	N.A.	N.A.
V_{CE} (max)	30 V	30 V	30 V	30 V	30 V	15 V
I_F (max)	60 mA	60 mA	100 mA	100 mA	50 mA	40 mA
CTR (min)	20%	300%	12.5%	12.5%	10%	0.5%
bandwidth	300 kHz	30 kHz	200 kHz	200 kHz	300 kHz	20 kHz
outline	Figure 6.6	Figure 6.7	Figure 6.8	Figure 6.9	Figure 6.10	Figure 6.11

Figure 6.12   *Typical parameter values of the* Figure 6.6 *to* 6.11 *devices*

stage and has a useful bandwidth of only 20 kHz. Even so, the device has a typical minimum CTR value of only 0.5% at a reflective range of 5 mm from a surface with a reflection efficiency of 90%, when the input LED is operated at its maximum current of 40 mA.

Usage notes

Optocouplers are very easy devices to use, with the input side being used in the manner of a normal LED and the output used in the manner of a normal phototransistor, as described in earlier chapters. The following notes give a summary of the salient usage points.

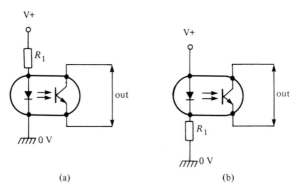

Figure 6.13   *The LED current must be limited via a series resistor, which can be connected to either the anode (a) or the cathode (b)*

The input current to the optocoupler LED must be limited via a series-connected external resistor which, as shown in *Figure 6.13*, can be connected on either the anode or the cathode side of the LED. If the LED is to be driven from an AC source, or there is a possibility of a reverse voltage being applied across the LED, the LED must be protected from reverse voltages via an external diode connected as shown in *Figure 6.14*.

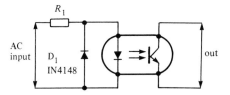

Figure 6.14 *The input LED can be protected against reverse voltages via an external diode*

The operating current of the phototransistor can be converted into a voltage by wiring an external resistor in series with the collector of the device. This resistor can in fact be connected to either the collector or the emitter of the phototransistor as shown in *Figure 6.15*. The greater the value of this resistor, the greater is the sensitivity of the circuit but the lower is its bandwidth.

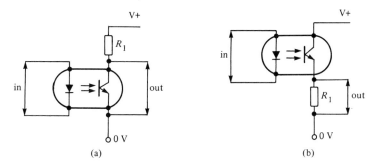

Figure 6.15 *An external output resistor, wired in series with the phototransistor, can be connected to either the collector (a) or emitter (b)*

In normal use, the phototransistor is used with its base terminal open circuit. If desired, however, the phototransistor can be converted into a photodiode by using the base terminal as shown in *Figure 6.16* and ignoring the emitter terminal (or shorting it to the base). This connection results in a greatly increased bandwidth (typically 30 MHz), but a greatly reduced CTR value (typically 0.2%).

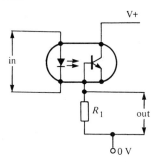

Figure 6.16   *If its base is available, the phototransistor can be made to function as a photodiode*

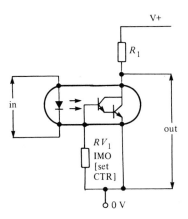

Figure 6.17   *The CTR values can be varied via* RV_1

Alternatively, the base terminal can be used to vary the CTR value of the optocoupler by wiring an external resistor (RV_1) between the base and emitter, as shown in the Darlington example of *Figure 6.17*. With RV_1 open circuit, the CTR value is that of a normal Darlington optocoupler (typically 300% minimum); with RV_1 short circuit, the CTR value is that of a diode-connected phototransistor (typically about 0.2%).

Digital interfacing

Optocoupler devices are ideally suited for use in digital interfacing applications in which the input and output circuits are driven by different power supplies. They can be used to interface digital ICs of the same family (TTL,

CMOS, etc.) or digital ICs of different families, or to interface the digital outputs of home computers, etc., to motors, relays and lamps, etc.

Figure 6.18 shows how to interface TTL circuits. The optocoupler LED and current-limiting resistor R_1 are connected between the 5 V positive supply rail and the output driving terminal of the TTL device (rather than between the TTL output and ground), because TTL outputs can usually sink fairly high current (typically 16 mA) but can source only a very low current (typically 400 μA).

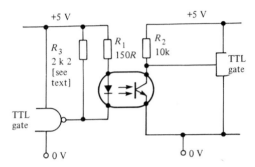

Figure 6.18   *TTL interface*

The open-circuit output voltage of a TTL IC falls to less than 0.4 V when in the logic-0 state, but may rise to only 2.4 V in the logic-1 state if the IC is not fitted with an internal pull-up resistor. In such a case the optocoupler LED current will not fall to zero when the TTL output is at logic-1. This snag can be overcome by fitting an external pull-up resistor (R_3) as shown in *Figure 6.18*.

The optocoupler phototransistor should be wired between the input and ground of the TTL IC as shown, since a TTL input needs to be pulled down to

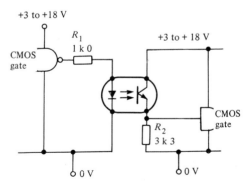

Figure 6.19   *CMOS interface*

below 800 mV at 1.6 mA to ensure correct logic-0 operation. Note that the *Figure 6.18* circuit provides non-inverting optocoupler action.

CMOS IC outputs can source or sink currents (up to several mA) with equal ease. Consequently, these devices can be interfaced by using a sink configuration similar to that of *Figure 6.18*, or they can use the source configuration shown in *Figure 6.19*. In either case, the R_2 value must be large enough to provide an output voltage swing that switches fully between the CMOS logic-0 and logic-1 states.

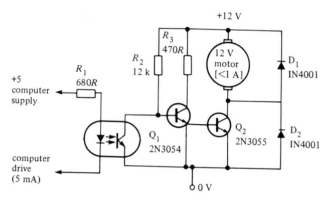

Figure 6.20   *Computer-driven DC motor*

Figure 6.20 shows how the optocoupler can be used to interface a computer's output signal (5 V, 5 mA) to a 12 V DC motor that draws an operating current of less than 1 A. With the computer output high, the optocoupler LED and phototransistor are both off, so the motor is driven on via Q_1 and Q_2. When the computer output goes low, the LED and phototransistor are driven on, so Q_1–Q_2 and the motor are cut off.

Analogue interfacing

An optocoupler can be used to interface analogue signals from one circuit to another by setting up a standing current through the LED and then modulating this current with the analogue signal. *Figure 6.21* shows this technique used to make an audio-coupling circuit.

Here, the op-amp is connected in the unity-gain voltage follower mode, with the optocoupler LED wired into its negative feedback loop so that the voltage across R_3 (and thus the current through the LED) precisely follows the voltage applied to the non-inverting input terminal (pin 3) of the op-amp. This terminal is DC biased at half-supply volts via the R_1–R_2 potential

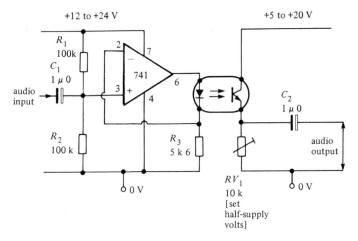

+12 to +24 V   +5 to +20 V

R_1
100k
C_1
$1 \mu 0$

audio
input

R_2
100 k

R_3
5 k 6

741

C_2
$1 \mu 0$

audio
output

RV_1
10 k
[set
half-supply
volts]

0 V   0 V

Figure 6.21   *Audio-coupling circuit*

divider, and can be AC-modulated by an audio signal applied via C_1. The quiescent LED current is set at 1 to 2 mA via R_3.

On the output side of the optocoupler, a quiescent current is set up (by the optocoupler action) in the phototransistor, and causes a quiescent voltage to be set up across RV_1, which should have its value adjusted to give a quiescent output value of half-supply voltage. The audio output signal appears across RV_1 and is decoupled via C_2.

Triac interfacing

An ideal application for the optocoupler is that of interfacing the output of a low-voltage control circuit (possibly with one side of its power supply grounded) to the input of a triac power-control circuit which is driven from the AC mains power lines, and which can be used to control the power feed to lamps, heaters, and motors. *Figures 6.22* and *6.23* show practical examples of such circuits; in these diagrams the figure in parenthesis show the component values that should be used if 115 V mains (rather than 230 V) supplies are used; actual triac types should be chosen to suit individual load/supply requirements.

The *Figure 6.22* circuit gives a non-synchronous switching action in which the triac's initial switch-on point is not synchronized to the mains voltage waveform. Here, R_2–D_1–ZD_1 and C_1 are used to develop a mains-derived 10 V DC supply, which can be fed to the triac gate via Q_1 and hence used to turn the triac on or off. Thus, when SW_1 is open the optocoupler is off, so zero base drive is applied to Q_1, and the triac and load are off. When SW_1 is closed, the opto-coupler drives Q_1 and connects the 10 V DC supply to the triac gate via R_3, thus applying full mains power to the load.

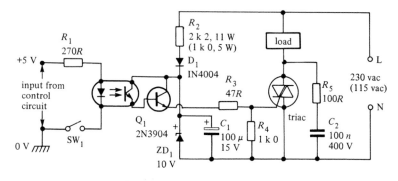

Figure 6.22   *Non-synchronous triac power switch with optocoupled input*

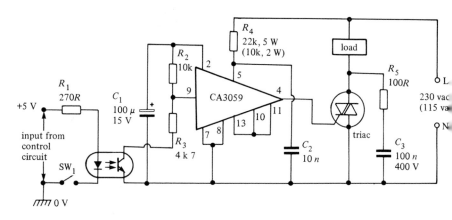

Figure 6.23   *Synchronous triac power switch with optocoupled input*

The *Figure 6.23* circuit uses a special zero-voltage switching IC in conjunction with an optocoupler, to give a synchronous power switching action in which gate current can be applied to the triac only when the instantaneous AC mains voltage is within a few volts of the zero cross-over value. This synchronous switching technique enables power loads to be switched on without generating sudden power surges (and thus RFI) in the power lines.

Opto-triacs

The silicon controlled rectifier, or SCR, is a fairly simple device that can easily be simulated with a pair of silicon transistors. Like the transistor, the SCR is inherently photosensitive, and is in fact available in a phototriggered form. Similarly, the triac is a fairly simple development of the SCR, and is also readily available in phototriggered form.

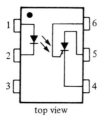

top view

Figure 6.24   *Typical optocoupled SCR*

From the above, it can be seen that an optocoupled SCR can be made by mounting an SCR and a LED in a single package, and that an optocoupled triac can be made by mounting a triac and a LED in a single package. Such devices are in fact available, and *Figures 6.24* and *6.25* show their typical outlines (they are usually mounted in six-pin DIL packages); *Figure 6.26* lists the typical parameters of these devices.

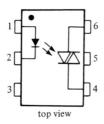

top view

Figure 6.25   *Typical optocoupled triac*

Note that the above devices have rather limited output-current ratings, actual r.m.s. values being (in the examples shown) 300 mA for the SCR and 100 mA for the triac. Like all other semiconductor devices, however, the surge-current ratings of these devices are far greater than the r.m.s. values. In the case of the SCR the surge current rating is 5 A, but this applies to a pulse of 100 μS width and to a duty cycle of less than 1%. In the case of the triac the surge rating is 1.2 A, and this applies to a 10 μS pulse width and a duty cycle of 10% maximum.

parameter	opto-coupled SCR	opto-coupled triac
LED characteristic		
I_F (max)	60 mA	50 mA
SCR/triac characteristics		
V_{MAX}	400 V	400 V
I_{MAX} (r.m.s.)	300 mA	100 mA
I_{SURGE} (see text)	5 A	1.2 A
coupled characteristics		
isolating voltage	1.5 kV	1.5 kV
input current needed to	5 mA typ	5 mA typ
trigger output device	20 mA max	20 mA max

Figure 6.26   *Typical characteristics of optocoupled SCRs/triacs*

Applications

Optocoupled SCRs and triacs are easy devices to use; the input LED is driven in the manner of a normal optocoupler LED, and the SCR/triac is used in the manner of a standard SCR/triac of limited current-handling capacity. *Figures 6.27* to *6.29* show some practical ways of using an optocoupled triac. In all cases, R_1 should be chosen to pass an LED on current of at least 20 mA.

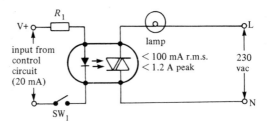

Figure 6.27   *Low-power lamp control*

In *Figure 6.27* the triac is used to directly activate a mains-powered filament lamp, which should have an r.m.s. rating of less than 100 mA and a peak inrush current rating of less than 1.2 A.

Figure 6.28 shows how the optocoupled triac can be used to activate a slave triac, and thereby activate a load of any desired power rating. This circuit is

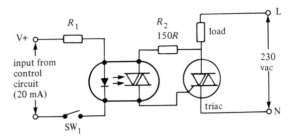

Figure 6.28   *High-power control via a triac slave*

suitable for use only with non-inductive loads such as lamps and heating elements.

Finally, *Figure 6.29* shows how the above circuit can be modified for use with inductive loads such as motors. The R_2–C_1–R_3 network provides a degree of phase-shift to the triac gate-drive network, to ensure correct triac triggering action, and R_4–C_2 form a snubber network, to suppress rate effects.

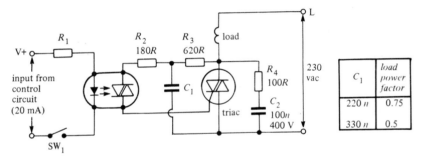

Figure 6.29   *Driving an inductive load*

7 Brightness-control techniques

In earlier chapters we have discussed light-generating devices such as tungsten filament lamps and light-emitting diodes (LEDs), etc. In most practical applications these devices are used in the simple ON/OFF mode in which they are always either at full brilliance or are fully off. In fact, however, the brilliance of these devices is fully variable between these two extremes; brightness-control techniques form the subject of this chapter.

Lamp-control basics

The brilliance of a DC-powered tungsten filament lamp can be varied in any one of three basic ways. The simplest way is to wire a rheostat and a ganged switch in series with the lamp as shown in *Figure 7.1*. Here, if RV_1 has a maximum resistance value double that of the 'hot' resistance value of the lamp, then RV_1 will enable the lamp power dissipation (and thus its brilliance) to be varied over an approximately 12:1 range, as explained below.

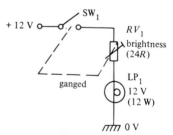

Figure 7.1   *Rheostat brightness control circuit*

A tungsten lamp has a positive temperature coefficient of resistance, causing the lamp resistance to increase with operating temperature. Thus, the hot or operating resistance (at which the filament is white hot) of a 12 V 12 W lamp is 12 Ω, but the cold resistance of the same lamp is typically only one quarter of this value (3 Ω), resulting in a fairly high 'inrush' lamp current at the moment of initial switch-on. The 'warm' resistance of the same lamp (at which the filament is a dull red) is typically about 6 Ω.

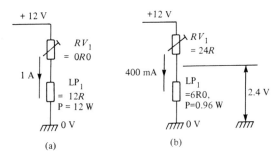

(a)   (b)

Figure 7.2   *Equivalent of the* Figure 7.1 *circuit at (a) maximum and (b) minimum brilliance levels*

From the above it can be seen that the operation of the *Figure 7.1* rheostat circuit is moderately complex, as is made clear in *Figure 7.2*. Thus, when RV_1 is set to the maximum brilliance (0 Ω) position the full supply voltage is applied to the lamp, which presents a resistance of 12 Ω and thus has a power consumption of 12 W, as shown in *Figure 7.2(a)*.

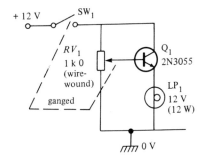

Figure 7.3   *Variable-voltage brightness-control circuit*

When RV_1 is set to the minimum brilliance (maximum resistance) position, however, RV_1 has a resistance of 24 Ω and the lamp presents a resistance of only 6 Ω, as shown in *Figure 7.2(b)*. Consequently, only 2.4 V are developed across the lamp, which thus consumes only 960 mW of power and produces very little light output. RV_1 thus allows the lamp brilliance to be varied over a wide range.

A major disadvantage of the *Figure 7.1* circuit is that a great deal of power is wasted in RV_1, which must have a substantial power rating and be capable of handling the cold currents of the lamp. *Figure 7.3* shows an alternative brilliance-control circuit, which dissipates negligible power in RV_1.

In *Figure 7.3*, RV_1 acts as a variable potential divider which applies an input voltage to the base of emitter follower Q_1, which buffers (power boosts) this voltage and applies it to the lamp. RV_1 thus enables the lamp voltage (and thus its brilliance) to be fully varied from zero to maximum. A disadvantage of this circuit is that Q_1 needs a large power rating and must be capable of handling the cold currents of the lamp.

Switched-mode control

The third and most sophisticated way of controlling the brilliance of a DC-powered lamp is the so-called switched-mode method, which is shown in basic form in *Figure 7.4*. Here, an electronic switch (SW_1) is wired in series with the lamp and can be opened and closed via a pulse-generator waveform. When this pulse is high, SW_1 is closed and power is fed to the lamp; when the pulse is low SW_1 is open, and power is not fed to the lamp.

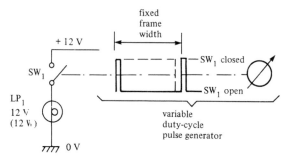

Figure 7.4   *Basic switched-mode brightness-control circuit*

The important thing to note about the *Figure 7.4* pulse generator is that it generates a waveform with a fixed frame width but with a variable mark/space (M/S) (ON/OFF) ratio or duty cycle, thereby enabling the MEAN lamp voltage to be varied. Typically, the M/S ratio is fully variable from 1:20 to 20:1, enabling the mean lamp voltage to be varied from 5% to 95% of the supply-voltage value.

Because of the inherently long thermal time constant of a tungsten lamp, its brilliance responds relatively slowly to rapid changes in input power.

Consequently, if the frame width of the *Figure 7.4* waveform generator is less than roughly 100 ms (i.e., the repetition frequency is greater than 10 Hz), the lamp will show no sign of flicker, and the lamp brilliance can be varied by altering the M/S ratio.

Thus, if the M/S ratio of the *Figure 7.4* circuit is set at 20:1, the mean lamp voltage is 11.4 V and the consequently hot lamp consumes 10.83 W. Alternatively, with the M/S ratio set at 1:20, the mean lamp voltage is only 600 mV, so the lamp is virtually cold and consumes a mere 120 mW. The lamp power consumption can thus be varied over a 90:1 range via the M/S-ratio control. Note, however, that this wide range of power control is obtained with virtually zero power loss within the system, since power is actually controlled by SW_1, which is always either fully on or fully off. The switched-mode control system is thus exceptionally efficient.

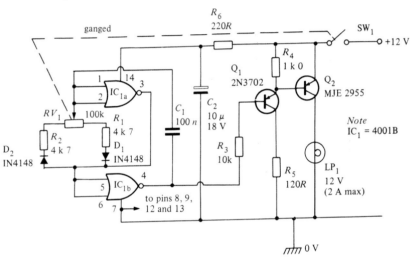

Figure 7.5 *Switched-mode DC lamp dimmer (− ve ground version)*

Figure 7.5 shows the practical circuit of a switched-mode DC lamp dimmer or brilliance control that is designed for use with a 12 V lamp with a maximum power rating of 24 W and enables the lamp's light intensity to be smoothly varied from zero to full brilliance via 100 k variable resistor RV_1. The circuit operates as follows.

In *Figure 7.5*, IC_{1a} and IC_{1b} (comprising one half of a 4001B CMOS quad two-input NOR gate) are wired as an astable multivibrator or square wave generator in which half of the waveform is generated via C_1–D_1–R_1 and the right-hand part of RV_1, and the other half is generated via C_1–D_2–R_2 and the left-hand part of RV_1, thus enabling the M/S-ratio to be varied via RV_1.

Thus, when SW_1 is closed the astable multi operates and feeds a switching waveform to the lamp via Q_1 and Q_2. The astable operates at a fixed frequency of about 100 Hz, but its M/S-ratio is fully variable from 1:20 to 20:1 via RV_1, thus enabling the mean lamp power to be varied over a 90:1 range. Note that ON/OFF switch SW_1 is ganged to RV_1, so that the circuit can be switched fully off by turning the RV_1 brilliance control fully anticlockwise.

The *Figure 7.5* lamp-dimmer circuit can be used to control the brilliance of virtually any low power (up to 24 W) filament lamps that are powered by 12 V DC supplies. Note, however, that if it is used to control car lights it can only be used in vehicles in which the free ends of the lamps go to the + ve supply line via control switches, as is normal in vehicles fitted with negative ground electrical systems (in which the − ve battery terminal goes to the chassis of the vehicle).

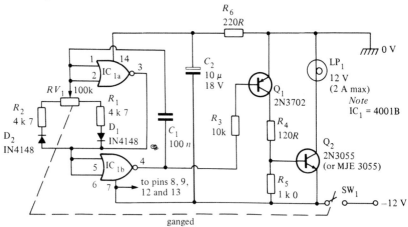

Figure 7.6   *Switched-mode DC lamp dimmer (+ ve ground version)*

The alternative circuit of *Figure 7.6* can be used to control the lights of vehicles in which the free end of the lamp goes to the − ve supply line via control switches, as normally occurs in vehicles fitted with positive ground electrical systems. Note in both the *Figure 7.5* and *Figure 7.6* circuits that R_6–C_2 are used to protect IC_1 against damage from high-voltage transients that may occur on the vehicle's supply lines.

LED brightness control

The switched-mode technique can also be used to control the brilliance of solid-state light-emitting devices such as LEDs and seven-segment LED displays. These devices, however, give an instant response to changes in input power level, so the design technique must rely on the natural integrating action of the human eye to ensure a flicker-free brightness-control action.

The natural action of the human eye is such that it ignores the instantaneous values of rapidly repeating changes in light level if these changes occur at a frequency in excess of about 40 Hz, and sees these changes in terms of the mean value of light intensity instead. Thus, in LED brightness-control circuits, the variable M/S-ratio generator normally operates in the range 50 to 100 Hz.

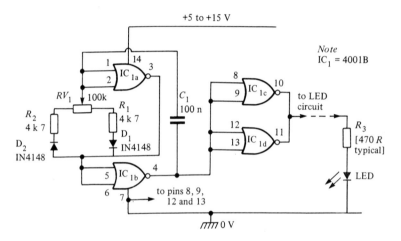

Figure 7.7   *Switched-mode LED brightness-control circuit*

Figure 7.7 shows a practical example of a LED brightness control circuit, designed around a single 4001B CMOS IC. Here, IC_{1a}–IC_{1b} are wired as a 100 Hz astable multi, with M/S-ratio variable from 1:20 to 20:1 via RV_1 and IC_{1c} and IC_{1d} are connected in parallel to provide a medium-current (15 to 20 mA) buffered drive to the LED via current-limiter R_3.

Figure 7.8 shows how to apply switched-mode brightness control to a common-cathode seven-segment LED display. Here, IC_{1a}-IC_{1b} are again wired as a 100 Hz astable with variable M/S-ratio, but in this case the output waveform is fed (via R_3) to the base of Q_1, which functions as a medium-power switch that is wired in series with the common cathode terminal of the display.

Figure 7.9 shows how to modify the above circuit for use with a common-anode display. Here, IC_{1c} is used as an inverting buffer that connects the astable output signal to the base of *pnp* transistor Q_1.

In practice, many seven-segment LED driver ICs have a blanking terminal

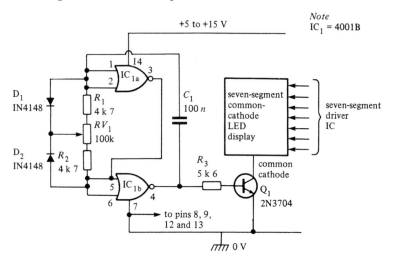

Figure 7.8   *Common-cathode LED brightness-control circuit*

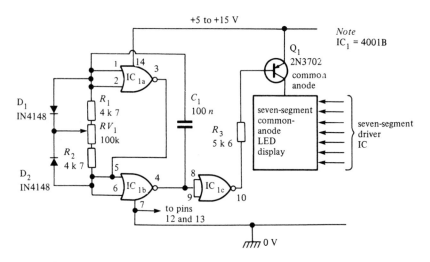

Figure 7.9   *Common-anode LED brightness-control circuit*

(enabling the display to be turned on and off) which can be used to apply switched-mode brightness control to the LED display device. This terminal is usually designated BL or BI. The TTL 7447 and 7448 range of decoder/driver ICs have such a terminal, as also does the CMOS 4511B latch/decoder/driver.

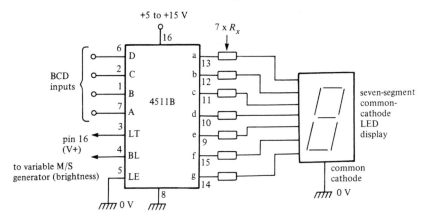

Figure 7.10   *LED brightness control via a 4511B driver IC*

In the latter case, the display is blanked when the BL terminal is low, and is active when the terminal is high. *Figure 7.10* shows how to connect this IC to give brightness control via an external variable M/S-ratio generator (such as is used in *Figures 7.5* to *7.9*).

AC lamp-control basics

The brilliance of an AC-powered lamp can be controlled by using a triac and a variable phase-delay network to vary the power feed to the lamp, as shown in the basic phase-triggered system of *Figure 7.11*. The triac is a bidirectional (AC) solid-state self-latching power switch that can be turned on by applying a brief trigger pulse to its gate, but which turns off again automatically at the end of each power half-cycle as its main-terminal currents fall to near-zero.

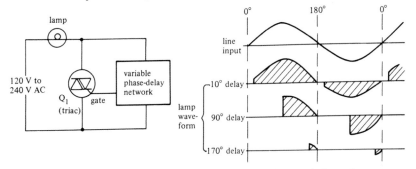

Figure 7.11   *Basic phase-triggered AC brightness-control circuit and waveforms*

Thus, in *Figure 7.11*, the triac is triggered via variable phase-delay network that is interposed between the AC power line and the triac gate. Hence, if the triac is triggered 10° after the start of each half-cycle, almost the full available power is fed to the lamp load. If the triac is triggered 90° after the start of each half cycle, only half of the available line power is fed to the load. Finally, if the triac is triggered 170° after the start of each half-cycle (e.g., 10° before the end of each half-cycle), only a very small part of the available power is fed to the load.

The three most popular methods of obtaining variable phase-delay triggering are to use either a line-synchonized UJT (unijunction transistor), or a special-purpose IC, or to use a diac and an *R–C* network in the basic configuration shown in *Figure 7.12*. The diac is a bilateral threshold switch which, when connected across a voltage source, presents a high impedance until the applied voltage rises to about 35 V, at which point the device switches into a low-impedance state and remains there until the applied voltage falls to about 30 V, at which point it reverts back to the high-impedance mode.

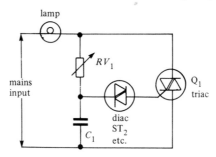

Figure 7.12   *Basic diac-type variable phase-delay lamp dimmer circuit*

Thus, in *Figure 7.12*, in each mains half-cycle the RV_1–C_1 network applies a variable phase-delayed version of the mains waveform to the triac gate via the diac, and each time that the C_1 voltage rises to 35 V the diac fires and delivers a trigger pulse to the triac gate, thus turning the triac on and simultaneously applying power to the lamp load and removing the drive from the RV_1–C_1 network. The mean power to the load (integrated over a full half-cycle period) is thus fully variable from near-zero to maximum via RV_1.

Radio-frequency interference

Note from the *Figure 7.11* waveforms that, each time the triac is gated on, the load current transitions abruptly (in a few microseconds) from zero to a value determined by the lamp resistance and the value of instantaneous mains voltage. These transitions inevitably generate radio-frequency interference (RFI). The RFI is greatest when the triac is triggered at 90°, and is least when the triac is triggered close to the 0° and 180° zero crossing points of the mains waveform.

In the lamp-dimmer brightness-control circuits, where there may be considerable lengths of mains cable between the triac and the lamp load, this RFI may be offensive. Consequently, in practical lamp dimmers, the circuit is usually provided with an *L–C* RFI-suppression network, as shown in the circuit of *Figure 7.13*; this circuit also shows how ON/OFF switch SW_1 can be ganged to brightness control pot RV_1.

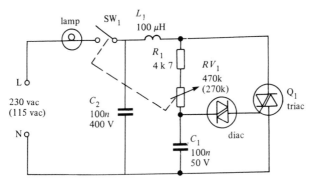

Figure 7.13   *Practical circuit of a simple lamp dimmer with RFI suppression*

Note in *Figure 7.13* (and all other triac circuits shown in this chapter) that the diac can be virtually any commercial type (ST2, etc.), that the triac type should be chosen to suit the mains voltage and lamp-load values, and that the values shown in brackets are applicable to 115 V (rather than 230 V) mains operation.

Back-lash reduction

The simple *Figure 7.13* circuit makes a useful lamp dimmer but has one annoying characteristic, in that RV_1 has considerable hysteresis or backlash. If, for example, the lamp finally goes fully off when RV_1 is increased to 470 k (in the 230 V circuit), it may not start to go on again until RV_1 is reduced to 400 k, and it then suddenly burns at a fairly high brightness level. The cause of this characteristic is as follows.

The basic action of the *Figure 7.13* circuit is such that, in the first part of each mains half-cycle, C_1 charges via RV_1–R_1 and the lamp, etc., until C_1 charges to 35 V, at which point the diac suddenly fires and starts to partially

discharge C_1 into the gate of the triac; as the triac turns on it switches the
remaining part of the half-cycle to the lamp and simultaneously removes the
mains drive from R_1–RV_1. This switching action only takes 2 μS or so, but in
this brief period the diac is able (in the *Figure 7.13* circuit) to remove
substantial charge (typically about 5 V) from C_1, and thus upsets the timing of
the following half-cycle, thereby causing the annoying backlash
characteristic.

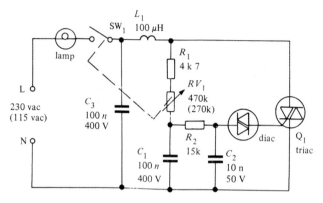

Figure 7.14   *Improved lamp dimmer with gate slaving*

One easy way to reduce this backlash is to simply wire a current-limiting
resistor (47R to 120R) in series with the diac, to reduce the amount of C_1
voltage change that takes place in the 2 μS triac-switch-on period. Another
way is to use the gate-slaving technique shown in *Figure 7.14*.

The *Figure 7.14* circuit is similar to that of *Figure 7.13*, except that the
charge of C_1 is coupled to slave capacitor C_2 via the relatively high resistance

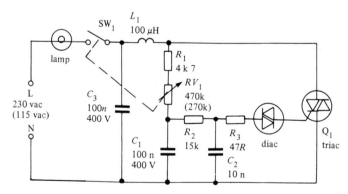

Figure 7.15   *Minimum-backlash lamp dimmer*

of R_2. C_1 thus changes to a slightly higher voltage than C_2, and C_2 fires the diac once its voltage reaches 35 V. Once the diac has fired it reduces the C_2 potential briefly to 30 V, but has little influence on the voltage value of the C_1 main-timing capacitor; the circuit backlash is thus reduced. The backlash can be reduced even more by wiring a current-limiting resistor in series with the diac (as described above), to reduce the magnitude of the C_2 discharge voltage, as shown in *Figure 7.15*.

Unijunction transistor-triggering

A lamp dimmer with absolutely zero backlash can be made by using a mains-synchronized variable-delay unijunction transistor (UJT) circuit to trigger the lamp-driving triac in each mains voltage half-cycle. *Figure 7.16* shows such a circuit. Here, the UJT is powered from a 12 V DC supply derived from the AC power line via R_1–D_1–ZD_1 and C_1. The UJT is synchronized to the mains via the Q_2–Q_3–Q_4 zero-crossing detector network, the action being such that Q_4 is turned on (applying power to the UJT circuit) via Q_2 or Q_3 at all times other than when the instantaneous mains voltage is close to the zero-crossover point at the end or the start of each mains half-cycle.

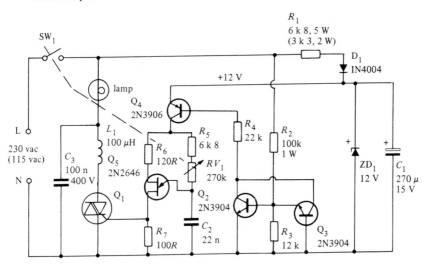

Figure 7.16   *UJT-triggered zero-backlash lamp dimmer*

Thus, shortly after the start of each half-cycle, power is applied to the UJT circuit via Q_4, initiating the start of the UJT timing cycle, and a short time later (determined by R_5–RV_1–C_2) the UJT (Q_5) delivers a trigger pulse to the triac gate, driving the triac on and connecting power to the lamp load for the remaining part of the half-cycle. The triac and the UJT circuit automatically reset at the end of each half-cycle, and a new sequence then begins.

The *Figure 7.16* circuit generates absolutely zero control backlash, and can be usefully modified for use in a variety of non-standard applications. *Figure 7.17*, for example, shows a circuit that can be fitted in place of the existing UJT network to modify *Figure 7.16* so that it acts as a slow-start lamp dimmer that simply causes the lamp brilliance to rise slowly from zero to maximum when first turned on, taking about two seconds to reach full brilliance. The circuit is intended to eliminate high turn-on inrush currents and to thus extend the lamp life. Circuit operation is as follows.

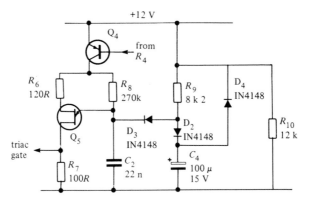

Figure 7.17   *Slow-start lamp-control circuit (for use with* Figure 7.16*)*

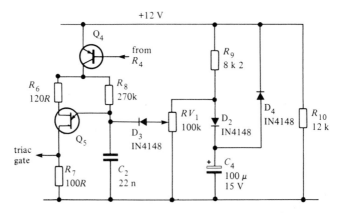

Figure 7.18   *Combined lamp dimmer and slow-start circuit (for use with* Figure 7.16*)*

When power is first applied to the circuit, C_4 is fully discharged and acts like a short circuit, so C_2 charges via high-value resistor R_8 only; the UJT thus generates a long delay under this condition, so the triac is triggered late in each half-cycle and the lamp burns at low brilliance. As time passes, C_4 slowly charges up via R_9, enabling the C_2 charge to be supplemented via R_9–D_3, thereby progressively reducing the UJT time constant and increasing the lamp brilliance until, when C_4 is fully charged (after roughly two seconds), full brilliance is reached.

Figure 7.18 shows how the above circuit can be further modified so that it acts as a combined lamp dimmer (via RV_1) and a slow-start circuit. Note in these two circuits that diode D_2 prevents C_4 from discharging into the UJT each time Q_5 fires, and D_4 automatically discharges C_4 via R_{10} and thus resets the network when the circuit is turned off.

S566B circuits

The brilliance of an AC-powered lamp can also be controlled with the aid of a dedicated phase-delaying IC such as the S566B, which is manufactured by Siemens. *Figure 7.19* shows the outline and pin designations of this device, which is housed in an eight-pin DIL package.

The Siemens S566B is a rather sophisticated IC which can be used to make a smart lamp dimmer that is controlled via either touch-sensing pads, or electromechanical push-button switches, or from external circuitry via an optocoupler device. The action of this chip, which feeds a phase-delayed trigger output to an external triac, is such that it alternately ramps up (increases brilliance) or ramps down (decreases brilliance) on alternate operations of the touch or push-button inputs, but remembers and holds brilliance levels when the input is released.

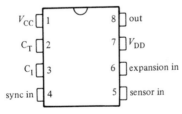

Figure 7.19 *Outline and pin notations of the S566B touch-sensitive lamp dimmer IC*

The IC actually incorporates touch conditioning circuitry, such that a very brief touch or push input causes the lamp to change state (from OFF to the remembered ON state, or vice versa), but a sustained (greater than 400 mS) input causes the IC to go into the ramping mode, in which the lamp power slowly ramps up from 3% to 97% of maximum and then down to 3% again, and so on, until the input is released, at which point the prevailing brilliance level is held and 'remembered'.

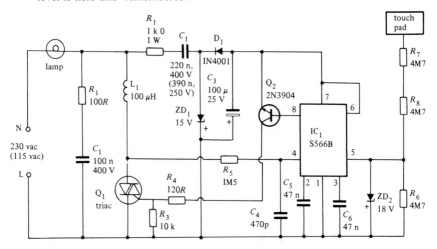

Figure 7.20 *Smart lamp dimmer, with touch-sensitive control*

Figure 7.20 shows how to connect the S566B and an external triac as a smart touch-controlled lamp dimmer. The circuit operates on the inductive pick-up principle, in which the human body picks up radiated low-frequency power-line signals when in the presence of mains-powered equipment, and these signals are detected when the touch pad is contacted. The touch pads can be simple strips or button pads of conductive material, and must be placed close

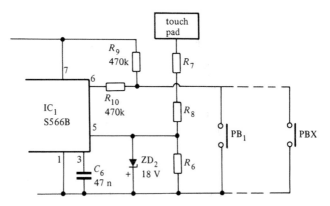

Figure 7.21 *Push-button control applied to the* Figure 7.20 *circuit*

to the IC (to avoid unwanted pick up); the operator is effectively and safely insulated from the mains voltage via the R_7–R_8 current-limiting resistors. For satisfactory touch operation, it is vital that the AC power lines be connected as shown, with the live or hot lead to pin 1 of the IC, and the neutral line to the lamp.

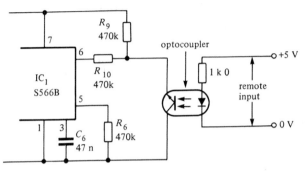

Figure 7.22   *Optocoupler operation of the S566B lamp dimmer circuit*

The touch pad must be placed close to the IC, to avoid unwanted pick-up signals. If multi-input operation is required, this can be obtained via push-button switches by modifying the circuit as shown in *Figure 7.21*. Here, the pin 6 to pin 7 connection is broken and replaced by the R_9–R_{10} divider. The pushbutton control switch (PB1) is connected between the R_9–R_{10} junction and pin 1 of the IC. Any desired number of push-button control switches can be connected in parallel. If the touch control facility is not needed, R_7–R_8 and ZD_2 can be eliminated and R_6 can be reduced to 470 k.

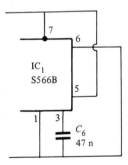

Figure 7.23   *Continuously-ramping version of the* Figure 7.20 *lamp dimmer*

Figure 7.22 shows how to modify the lamp dimmer so that it can be controlled by an external circuit via an optocoupler, without the use of the touch control facility.

Finally, to complete our look at the S566B IC, *Figure 7.23* shows how to modify the *Figure 7.20* circuit so that the lamp brightness level continuously ramps up and down from zero to maximum, and vice versa; this type of action is useful in advertising displays and burglar alarms, etc. Here, pin 5 of the IC is simply wired to pin 7, and pin 6 is wired to pin 1.

Infra-red LED pulsing

Infra-red (IR) LEDs are often used in remote-control systems. Here, the IR LED is used to transmit a coded invisible-light signal, which is detected by a matching infra-red diode (and subsequently decoded) in a receiver system some distance away.

To give an adequate remote-control range (up to 10 m), the IR LED must pass ON currents of several hundred milliamps, but for practical reasons the complete transmitter must be small enough to fit into the hand, must be self-powered via an inexpensive battery, and must be capable of giving many hours of continuous control operation before needing battery replacement. These conflicting requirements can be met by using the basic circuit of *Figure 7.24*, which is driven via the waveforms shown in *Figure 7.25*.

The coded IR-transmitter signal comprises 1 mS bursts of 20 kHz pulses,

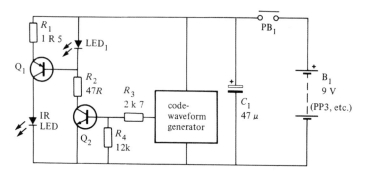

Figure 7.24   *Basic IR remote-control transmitter*

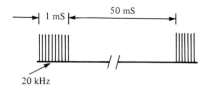

Figure 7.25   *Typical code waveform of the* Figure 7.24 *circuit*

repeated at 51 mS timebase intervals, i.e., at a 1 : 50 M/S-ratio. The transmitter generates peak IR LED currents of about 600 mA, giving a mean current of 300 mA during the 'mark' part of each transmitting cycle, but a mean of only 6 mA when averaged over the complete 51 mS timebase period.

In the actual transmitter, the coded waveform is fed to the base of Q_2 via R_3. When the waveform is high, Q_2 is driven to saturation, driving on Q_1 and LED_1 and feeding roughly 600 mA into the IR LED via R_1; when the waveform is low, zero current feeds into the IR LED. Note the capacitor C_1 acts as a low-impedance energy-storage unit and provided the required high drive currents to the IR LED; these currents could not be provided by battery B_1 alone. Some practical examples of complete IR remote-control systems are shown in Chapter 9.

8 Light-beam alarms

In the survey of modern optoelectronic devices and techniques in Chapter 1, we briefly mentioned light-beam systems which can, among other things, be used as the basis of intrusion detectors and alarms. In this chapter we expand on this theme by showing ways of making practical infra-red invisible-light-beam alarms. We start off by looking at some basic principles.

Intrusion alarm basics

A simple invisible-light-beam intrusion detector or alarm system can be made by connecting an infra-red (IR) light transmitter and receiver as shown in *Figure 8.1*. Here, the transmitter feeds a coded signal (often a simple square wave) into an IR LED which has its output focused into a fairly narrow beam (via a moulded-in lens in the LED casing) that is aimed at a matching IR photodetector (phototransistor or photodiode) in the remotely placed

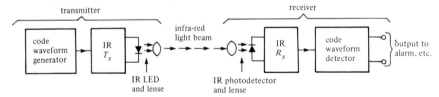

Figure 8.1 *Simple light-beam intrusion alarm/detector system*

receiver. The system action is such that the receiver output is 'off' when the light-beam reaches the receiver, but turns on and activates an external alarm, counter, or relay if the beam is interrupted by a person, animal, or object. This basic type of system can be designed to give an effective detection range of up to 30 m when used with additional optical focusing lenses, or up to 8 m without extra lenses.

144

The above system works on the pin-point line-of-sight principle and can be activated by any bigger-than-a-pin object that enters the line-of-sight between the transmitter and receiver lenses. Thus, a weakness of this simple system is that it can be false-triggered by a fly or moth (etc.) entering the beam or landing on one of the lenses. The improved dual-light-beam system of *Figure 8.2* does not suffer from this defect.

The *Figure 8.2* system is basically similar to that already described, but transmits the IR beam via two series-connected LEDs that are spaced about 75 mm apart, and receives the beam via two parallel-connected photo-detectors that are also spaced about 75 mm apart. Thus, each photodetector can detect the beam from either LED, and the receiver will thus activate only if BOTH beams are broken simultaneously, and this will normally only occur if a large (greater than 75 mm) object is placed within the composite beam. This system is thus virtually immune to false triggering by moths, etc.

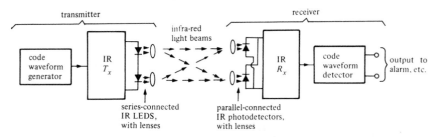

Figure 8.2   *Dual-light-beam intrusion alarm/detector system*

Note that, as well as giving excellent false-alarm immunity, the dual-light-beam system also gives (at any given LED drive-current value) double the effective detection range of the simple single-beam system (i.e., up to 16 m without additional lenses), since it has twice as much effective infra-red transmitter output power and twice the receiver sensitivity.

System waveforms

Infra-red beam systems are usually used in conditions in which high levels of ambient or background IR radiation (usually generated by heat sources such as radiators, tungsten lamps, and human bodies, etc.) already exist. To enable the systems to differentiate against this background radiation and give good effective detection ranges, the transmitter beams are invariably frequency modulated, and the receivers are fitted with matching frequency detectors. In practice, the transmitted beams invariably use either continuous-tone or tone-burst frequency modulation, as shown in *Figure 8.3*.

Infra-red LEDs and photodetectors are very fast acting devices, and the effective range of an IR beam system is thus determined by the *peak* current fed into the transmitting LED, rather than by its mean LED current. Thus, if the waveforms of *Figure 8.3* are used in transmitters giving peak LED currents of 100 mA, both systems will give the same effective operating range, but the *Figure 8.3(a)* continuous-tone transmitter will consume a mean LED current of 50 mA, while the tone-burst system of *Figure 8.3(b)* will consume a mean current of only 1 mA (but will require more complex circuit design).

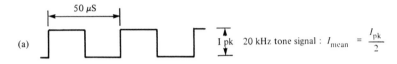

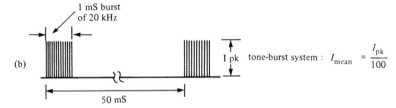

Figure 8.3   *Alternative types of IR light-beam code waveforms, with typical parameter values*

The operating parameters of the tone-burst waveform system require some consideration, since the system actually works on the sampling principle. For example, it is a fact that at normal walking speed a human takes about 200 mS to pass any given point, so a practical IR light-beam burglar alarm system does not need to be turned on continuously, but only needs to be turned on for brief 'sample' periods at repetition periods far less than 200 mS (at, say, 50 mS); the sample period should be short relative to repetition time, but long relative to the period of the tone frequency. Thus, a good compromise is to use a 20 kHz tone with a burst or sample period of 1 mS and a repetition time of 50 mS, as shown in *Figure 8.3(b)*.

System design

The first step in designing any electronics system is that of drawing up suitable block diagrams. *Figure 8.4* shows a suitable block diagram of a continuous-tone IR intrusion alarm/detector system, and that of *Figure 8.5* shows that of

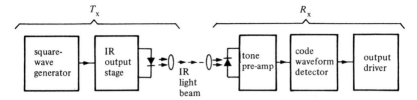

Figure 8.4 *Block diagram of continuous-tone IR light-beam intruder alarm/detector system*

a tone-burst system. Note that a number of blocks (such as the IR output stage, the tone pre-amp, and the output driver) are common to both systems. The continuous-tone system (*Figure 8.4*) is very simple, with the transmitter comprising nothing more than a square-wave generator driving an IR output stage, and the receiver comprising a matching tone pre-amplifier and code waveform detector, followed by an output driver stage that activates devices such as relays and alarms, etc.

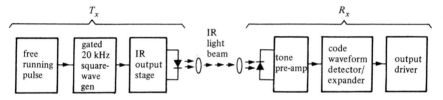

Figure 8.5 *Block diagram of tone-burst IR light-beam intruder alarm/detector system*

The tone-burst system is far more complex, with the transmitter comprising a free-running pulse generator (generating 1 mS pulses at 50 mS intervals) that drives a gated 20 kH$_2$ square-wave generator, which in turn drives the IR output stage, which finally generates the tone-burst IR light beam. In the receiver, the beam signals are picked up and passed through a matching pre-amplifier, and are then passed on to a code waveform detector/expander block, which ensures that the alarm does not activate during the 'blank' parts of the IR waveform. The output of the expander stage is fed to the output driver.

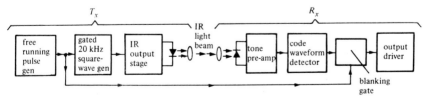

Figure 8.6 *Block diagram of alternative IR light-beam system*

Figure 8.6 shows an alternative version of the tone-burst system. This is similar to the above, except that a simple code waveform detector is used in the receiver section, and that a blanking gate is interposed between the detector and the output driver and is directly driven by the transmitter's pulse generator, to ensure that the alarm is not activated during the blank parts of the IR waveform.

Transmitter circuits

Figure 8.7 shows the practical circuit of a simple continuous-tone dual-light-beam IR transmitter. Here, a standard 555 timer IC is wired as an astable multivibrator that generates a non-symmetrical 20 kHz square-wave output that drives the two IR LEDs at peak output currents of about 400 mA via R_4 and Q_1 and the low source impedance of storage capacitor C_1. The timing action of this circuit is such that the ON period of the LEDs is controlled by C_2 and R_2, and the OFF period by C_2 and $(R_1 + R_2)$, i.e., so that the LEDs are ON for only about one eighth of each cycle; the circuit thus consumes a *mean* current of about 50 mA.

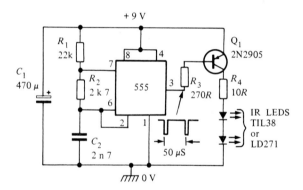

Figure 8.7   *Simple continuous-tone IR light-beam transmitter*

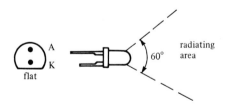

Figure 8.8   *Outline and connections of the LD271 and TIL38 IR LED*

The above circuit can use either TIL38 or LD271 (or similar) high-power IR LEDs. These devices can handle mean currents up to only 150 mA, but can handle brief repetitive peak currents several times greater than this value. *Figure 8.8* shows the outline and connections of these devices, which have a moulded-in lens that focuses the output into a radiating beam of about 60° width; at the edges of this beam the IR signal strength is half of that at the centre of the beam.

Minor weaknesses of the IR output stage (Q_1 and R_3–R_4) of the *Figure 8.7* circuit are that it has a very low input impedance (about 300 Ω), that it gives an inverting action (the LEDs are ON when the input is low), and that the LED output current varies with the circuit supply voltage. *Figure 8.9* shows an alternative universal IR transmitter output stage that suffers from none of these defects.

In *Figure 8.9*, the base drive current of output transistor Q_2 is derived from the collector of Q_1, which has an input impedance of about 5k0 (determined mainly by the R_1 value). Thus, when the input is low Q_1 is off, so Q_2 and the two IR LEDs are also off, but when the input is high Q_1 is driven to saturation via R_3, thus driving LED$_1$ (a standard red LED) and Q_2 and the two IR LEDs on.

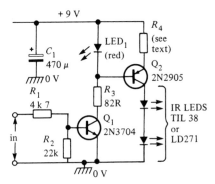

Figure 8.9 *Universal IR Tx output stage*

Note that under this latter condition about 1.8 V are developed across LED$_1$, and that about 0.6 V less than this (= 1.2 V) is thus developed across R_4. Consequently, since the R_4 voltage is determined by the Q_2 emitter current, and the Q_2 emitter and collector currents are virtually identical, it can be seen that Q_2 acts as a constant-current generator under this condition, and that the IR LED drive currents are virtually independent of variations in supply voltage. The peak LED drive current thus approximately equals $1.2/R_4$, and R_4 (in ohms) $= 1.2/I$, where I is the peak LED current in amps.

Figure 8.10 shows a 20 kHz square-wave generator (made from a 555 or

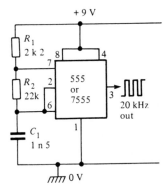

Figure 8.10   *20 kHz square-wave generator*

7555 timer IC) that can be added to the *Figure 8.9* output stage, to make a continuous-tone transmitter. In this case R_4 should be given a value of 8.2 Ω or greater, to limit peak LED currents to less than 150 mA.

Alternatively, *Figure 8.11* shows the circuit of a tone-burst generator (giving 1 mS bursts of 20 kHz at 50 mS intervals) that can be added to *Figure 8.9* to make a tone-burst transmitter. Here, two sections of a 40011B CMOS quad two-input NAND gate IC are wired as a non-symmetrical astable multivibrator producing 1 mS and 49 mS periods; this waveform is buffered by a third 4011B stage and used to gate a 20 kHz 555/7555 astable via D_2, and the output of the 555/7555 astable is then inverted via a fourth 4011B stage, ready for feeding to the transmitter output stage.

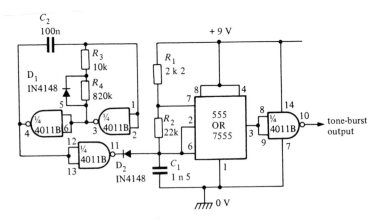

Figure 8.11   *Tone-burst (1 mS burst of 20 kHz at 50 ms intervals) waveform generator*

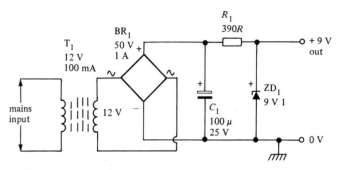

Figure 8.12 *Mains-powered 9 V supply*

Note when using the *Figure 8.11* circuit that R_4 in *Figure 8.9* can be given a value as low as $2.2\,\Omega$, to give peak output currents of about 550 mA, but that under this condition the transmitter will consume a mean current of little more than 6 mA; this current can be provided by either a battery or a mains-derived supply; a suitable mains-powered supply is shown in *Figure 8.12* (note that BR_1 is a bridge rectifier).

A receiver pre-amp

Figure 8.13 shows the practical circuit of a high-gain 20 kHz tone pre-amplifier designed for use in an infra-red receiver. Here, the two IR detectors are connected in parallel and wired in series with R_1, so that the detected IR signal is developed across this resistor. This signal is amplified by cascaded

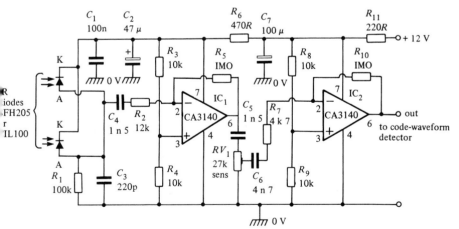

Figure 8.13 *Infra-red receiver pre-amplifier circuit*

op-amps IC_1 and IC_2, which can provide a maximum signal gain of about
$\times 17,680 (= \times 83$ via IC_1 and $\times 213$ via IC_2), but have gain fully variable via
RV_1. These two amplifier stages have their responses centred on 20 kHz, with
third-order low-frequency roll-off provided via C_4–C_5–C_6 and third-order
high-frequency roll-off provided by C_8 and the internal capacitors of the ICs.

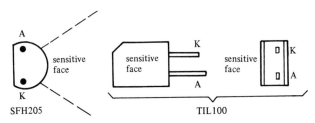

Figure 8.14 *Outline and connections of the SFH205 and TIL100 IR photodiodes*

The above circuit can be used with either SFH205 or TIL100 IR diodes;
these devices are housed in black infra-red transmissive mouldings which
reduce ambient white-light interference; *Figure 8.14* shows their outlines and
pin connections.

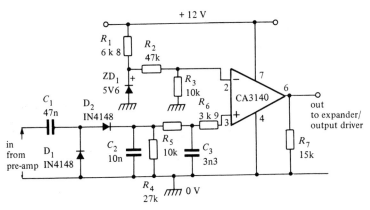

Figure 8.15 *Code waveform detector circuit*

The output of the *Figure 8.13* pre-amplifier can be taken from IC_2 and fed
directly to a suitable code-waveform detector circuit, such as that shown in
Figure 8.15. Note, however, that if the T_x–R_x system is to be used over ranges
less than 2 m or so the pre-amp output can be taken directly from IC_1, and all
the RV_1 and IC_2 circuitry can be omitted from the pre-amp design.

Code waveform detector

In the *Figure 8.15* code waveform detector circuit the 20 kHz tone waveforms (from the pre-amp output) are converted into DC via the D_1–D_2–C_2–R_5–C_3 network and fed (via R_6) to the non-inverting input of the op-amp voltage comparator, which has its inverting input connected to a thermally stable 1V0 DC reference point. The overall circuit action is such that the op-amp output is high (at almost full positive supply rail voltage) when a 20 kHz tone

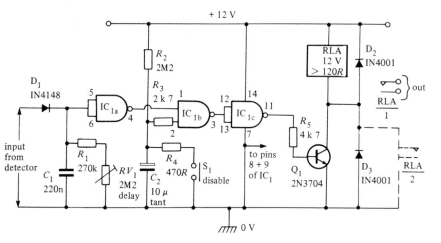

Figure 8.16 *Expander/output driver circuit*

input signal is present, and is low (at near-zero volts) when a tone input signal is absent; if the input signal is derived from a tone-burst system, the output follows the pulse-modulation envelope of the original transmitter signal. The detector output can be made to activate a relay in the absence of a beam signal by using the expander/output driver circuit of *Figure 8.16*.

Expander/output driver

The operating theory of the *Figure 8.16* circuit is fairly simple. When the input signal from the detector circuit switches high C_1 charges rapidly via D_1, but when the input switches low C_1 discharges slowly via R_1 and RV_1; C_1 thus provides a DC output voltage that is a 'time-expanded' version (with expansion presettable via RV_1) of the DC input voltage. This DC output voltage is buffered and inverted via IC_{1a} and used to activate relay RLA via Q_1 and an AND gate made from IC_{1b} and IC_{1c}.

Normally, the other (pin 2) input of this AND gate is biased high via R_2, and the circuit action is such that (when used in a complete IR light-beam system) the relay is off when the beam is present, but is driven on when the beam is absent for more than 100 mS or so. This action does not occur, however, when pin 2 of the AND gate is pulled low; under this condition the relay is effectively disabled.

The purpose of the R_2–C_2 network is to automatically disable the relay network via the AND gate (in the way described above) for several seconds after power is initially connected to the circuit or after DISABLE switch S_1 is briefly operated, so that the owner can safely pass through the beam without activating the relay. Note that the relay can be made self-latching, if required, by wiring normally-open relay contacts RLA/2 between Q_1 emitter and collector, as shown dotted in *Figure 8.16*.

A power supply

The circuits of *Figures 8.13, 8.15,* and *8.16* can be directly interconnected to make a complete infra-red light-beam receiver that can respond to either continuous-tone or tone-burst signals. Such a receiver should be powered via a regulated 12 V DC supply; *Figure 8.17* shows the circuit of a suitable mains-powered unit.

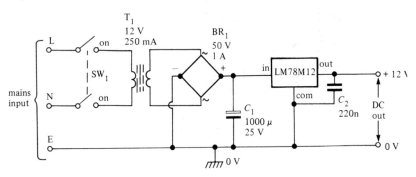

Figure 8.17   *Mains-powered regulated 12 V supply*

9 Remote control systems

In Chapter 1 we briefly mentioned light-beam systems which, among other things, can be used for remote control purposes. In this final chapter we expand on this theme by showing practical ways of making IR light beam remote-control units. We start off, however, by looking at some basic principles.

Remote-control basics

Infra-red beam systems can easily be designed to give very effective single- or multichannel remote-control operation; *Figure 9.1* illustrates the basic principle. Here, the hand-held control unit transmits a coded waveform via a

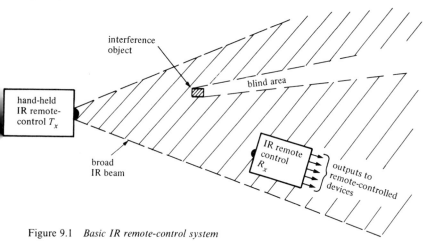

Figure 9.1   *Basic IR remote-control system*

broad infra-red beam, and this signal is detected and decoded in the remotely-placed receiver and thence used to activate external devices, etc., via the receiver outputs.

Note from *Figure 9.1* that the transmitter can remote-control a receiver that is placed anywhere within the active area of the IR beam, and that the transmitter and receiver do not need to be pointed directly at each other to effect operation but *must* be in line-of-sight contact; also note that an object placed within the beam can create a blind area in which line-of-sight contact cannot exist.

Code waveforms

Most modern IR remote-control systems give multi-channel operation, with each channel giving digital control of an individual function. The transmitter waveforms usually take the general form shown in *Figure 9.2*, which depicts those of a basic six-bit multichannel system. Here, the waveform comprises an 8 mS repeating frame of seven bits of pulse-coded information, with each bit modulated at about 30 kHz. The first bit has a fixed 1 mS duration and

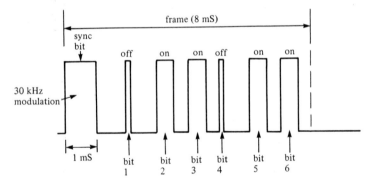

Figure 9.2   *Typical transmitter code waveform of a six-bit remote-control system*

provides frame synchronization for the decoder; the subsequent six data bits appear at 1mS intervals and each give an on/off form of control, with a less-than-0.25 mS pulse representing an OFF or logic-0 state and a greater-than-0.25 mS pulse representing an ON or logic-1 state. In practice, this six-bit code waveform can be used to give either six simultaneous channels or up to sixty-four non-simultaneous channels of remote control, as shown in *Figures 9.3* and *9.4*.

When the *Figure 9.2* waveforms are used to give simultaneous operation, each data bit controls a single channel, as shown in *Figure 9.3*. Channels 1 and

2 may thus each be used to give independent on/off switching functions, while channels 3 and 4 may be used to increase/decrease the output of a ramping volume control, and channels 5 and 6 may be used to similarly control the output of a ramping brilliance control, etc. Note that, within each transmitter frame, all six channels may be varied simultaneously.

channel numbers	channel state	decoded function
1	on off	switch A on switch A off
2	on off	switch B on switch B off
3	on off	volume increase
4	on off	volume decrease
5	on off	brilliance increase
6	on off	brilliance decrease

Figure 9.3   *Typical functions of a six-channel simultaneous remote-control system*

When the *Figure 9.2* waveforms are used to give non-simultaneous remote control operation the six data bits are used (within each frame) to generate a unique six-bit binary code, and each of these codes controls an individual remote-control channel; there are a total of sixty-four possible codes, and this system can thus be used to give up to sixty-four channels of remote control, as

channel number	six-bit code	decoded function
1	000 000	switch A on
2	000 001	switch A off
3	000 010	switch B on
4	000 011	switch B off
—	— — —	— — — —
61	111 100	volume increase
62	111 101	volume decrease
63	111 110	brilliance increase
64	111 111	brilliance decrease

Figure 9.4   *Typical functions of a sixty-four-channel non-simultaneous remote-control system*

shown in *Figure 9.4*. Thus, channels 1 to 4 may be used to control the on/off action of a pair of switches, and channels 61 to 64 may be used to control the levels of volume and brilliance; the remaining fifty-six channels may be used for a variety of other purposes! Note that, within each transmitter frame, only a single channel can be controlled at a time but that, since the frames are updated many times per second, this is usually only a minor disadvantage.

Block diagrams

Figure 9.5 shows the typical block diagram of a multichannel IR remote-control transmitter. This type of unit is usually fitted with a multifunction keyboard, which has its X and Y outputs repeatedly scanned via an encoder circuit that controls the input to a code waveform generator system. This latter unit generates the carrier wave signal (typically about 30 kHz) and the six-bit plus sync pulse repeating frame waveforms, which are then passed on to a standard infra-red transmitter output stage.

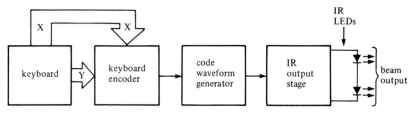

Figure 9.5   *Block diagram of a typical IR remote-control transmitter*

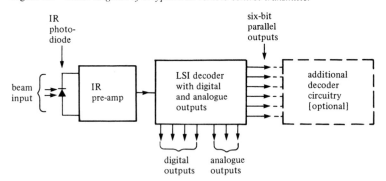

Figure 9.6   *Block diagram of a typical IR remote-control receiver*

In the receiver circuit (*Figure 9.6*) the detected IR signal is first fed to a fairly sophisticated pre-amplifier stage which provides very high gain for long-range operation but does not saturate if the transmitter is used near the receiver. The pre-amp output is fed to an LSI (large scale integration) decoder IC, which typically directly provides three or four digital outputs (simple on/off functions) and two or three analogue outputs (volume, brilliance, etc.), but also provides a six-bit output that is a parallel-coded version of the original six-bit serial code and can optionally be decoded via additional ICs to give more remote-control functions.

Practical systems

Simple single-channel IR remote-control systems can easily be built using discrete components such as bipolar transistors or VFETs, etc. Multichannel systems with up to six digital channels are only slightly more complicated, and can be built with the aid of simple CMOS ICs such as the 4017B, etc. If more than six channels are required, however, it is best to use dedicated LSI remote-control ICs for the purpose. Several manufacturers produce dedicated ICs of this type, the best known of these being the 490/922 thirty-two-channel range of devices from Plessey, and the IR60 sixty-channel range of devices from Siemens. We will examine both of these systems in the rest of this chapter.

The Plessey 490/922 system

System basics
The basic Plessey system is designed to give up to thirty-two-channels of non-simultaneous remote control via any of a variety of media, including ultrasound, IR beam, fibre optics, and direct cable link; in this book we will consider IR applications only.

The basic control system consists of one thirty-two-channel transmitter IC (the SL490), one infra-red pre-amplifier IC (the SL486), and one general-purpose receiver/decoder IC (the ML922) that provides three analogue and three digital outputs plus a four-bit binary output. Four additional receiver ICs (the ML926 to ML929) that each provide simple four-bit binary outputs (either latched or unlatched) are also available, enabling all thirty-two available channels to be accessed if required.

The SL490 transmitter IC
Figures 9.7 and *9.8* show the outline and block diagram of the SL490 transmitter IC, which is housed in an eighteen-pin DIL package. In use, a bank of up to thirty-two single-pole push-button switches (arranged in an eight-row by four-column matrix) form the transmitter keyboard. As each push button is operated the *encoder* and *code register* sections of the IC detect closure at a matrix crosspoint and generate a corresponding five-bit (up to

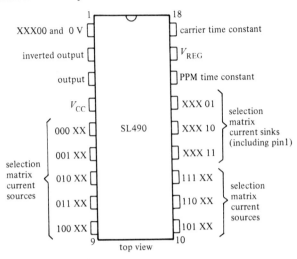

Figure 9.7   *Outline and pin notations of the SL490 transmitter IC*

thirty-two different codes) parallel code word, which is repeated at regular intervals until the push-button is released. The *multiplexer, three-bit counter,* and *pulse position modulator* sections of the IC (plus the *carrier oscillator* section, if required) then convert this parallel code word into a serial form which, with the addition of a suitable frame sync pulse, is passed on to the

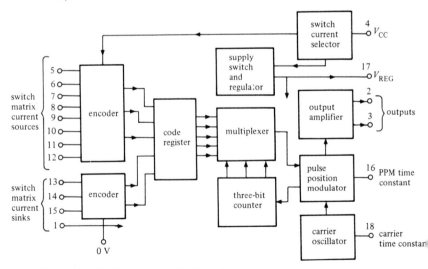

Figure 9.8   *SL490 transmitter IC block diagram*

output section for transmission via an external IR output stage. The pin 3 output signal takes the general form shown in *Figure 9.9*; an inverted version of this signal is available at pin 2.

The transmitted code waveform uses so-called pulse position modulation (PPM), in which each frame consists of six pulses of fixed length (L), but in which the position or spacing of each pulse depends on the code signal being transmitted. Specifically, a logic-1 code has a spacing of $6L$, a logic-0 code of $9L$, and a sync code of $18L$, as shown in *Figure 9.9*. The actual pulses are usually unmodulated, but can be modulated with a carrier signal if desired.

Figure 9.10 shows the practical circuit of a thirty-two-channel IR transmitter (unwanted channels can be deleted by simply omitting the

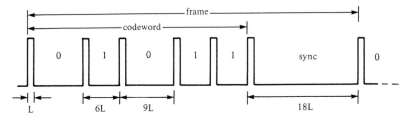

Figure 9.9   *Typical pin 3 output waveform of the SL490 transmitter IC*

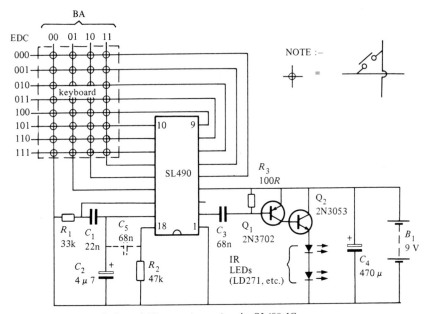

Figure 9.10   *Multichannel IR transmitter using the SL490 IC*

appropriate keyboard switches). Here, R_1–C_1 determine pulse length L ($L = 0.14\,C_1$. R_1 seconds) and give a period of about 1 mS with the component values shown, thus giving a nominal transmission rate of about 20 frames per second. In this specific application the pulse output waveforms are not carrier modulated, but C_3–R_3 cause the Q_1–Q_2 complementary output pair to conduct for about 15 μS at every negative leading edge of the PPM waveform, thus generating high peak currents in the two IR output LEDs. The circuit consumes a typical quiescent current of 8 μA in the standby mode.

Note in the above circuit that the pulse output signals can, if required, be carrier modulated by connecting C_5 as shown dotted in the diagram, in which case the carrier frequency $= 1/(C_5, R_2)$. If carrier modulation is used, the *Figure 9.10* circuit will need an alternative design of output stage.

SL486 pre-amp circuits

The SL486 is a four-stage high-gain IR receiver pre-amplifier IC featuring a fast-acting AGC circuit, a built-in voltage regulator, and a pulse-stretching output facility. *Figure 9.11* shows the outline and pin notations of this IC, which can be operated from supplies in the 4.5 to 18 V range.

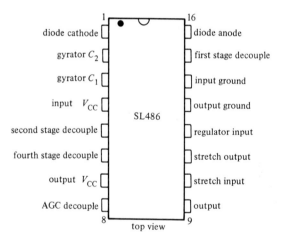

Figure 9.11   *Outline and pin notations of the SL486 IR pre-amp IC*

Figure 9.12 shows a minimum component application circuit for the SL486; this design is suitable for use with supplies in the 4.5 to 9 V range only, and gives an unstretched output; C_2, C_3 and C_5 provide stage decoupling, while C_4 controls the AGC time constant and C_1 influences the DC gyrator gain of the IC. Alternatively, *Figure 9.13* shows how the IC can be used with a 16 V supply that is common with the ML920 series of receiver ICs.

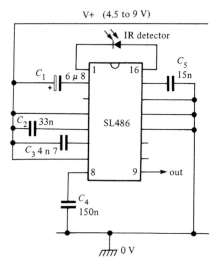

Figure 9.12   *SL486 application circuit for use with low voltage supplies*

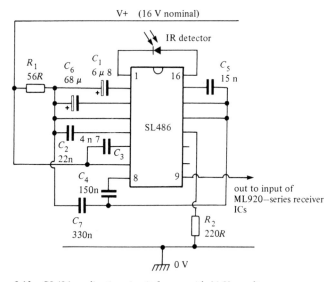

Figure 9.13   *SL486 application circuit for use with 16 V supplies*

Receiver IC circuits

Five different receiver ICs are available in the ML920 range of devices. Of these, the ML922 is the most versatile; it is housed in an eighteen-pin package (see *Figure 9.14*), provides three analogue and three digital outputs plus a four-bit binary output, and can respond to twenty-one different codewords.

The ML926 to ML929 range of ICs are less versatile; they are housed in eight-pin packages (see *Figure 9.15*) and provide simple four-bit binary

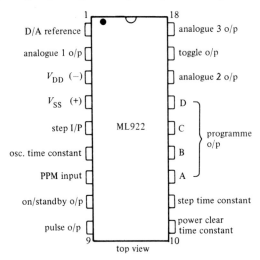

Figure 9.14   *Outline and pin notations of the ML922 receiver IC*

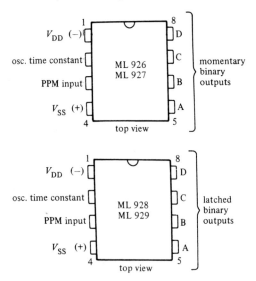

Figure 9.15   *Outline and notations of the ML926 to ML929 receiver ICs*

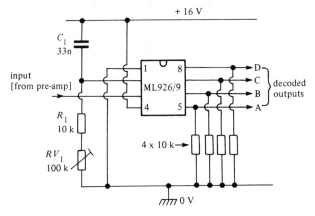

Figure 9.16   *Simple IR receiver circuit*

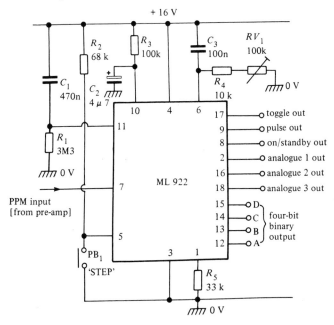

Figure 9.17   *Versatile IR receiver circuit*

outputs. The ML926 and ML927 give unlatched outputs and each respond to fifteen codewords (words 00001 to 01111 in the case of the ML926, and words 10001 to 11111 in the case of ML927). The ML928 and ML929 give latching outputs and each respond to sixteen codewords (words 00000 to 01111 in the case of the ML928, and words 10000 to 11111 in the case of the ML929). All five ICs are intended to operate from 16 V nominal supplies; actual limits are 14 V to 18 V for the ML922, and 12 V to 18 V for the ML926 to ML929 ICs.

All five ICs operate in the same basic way. They each incorporate an on-chip timing oscillator, which must be adjusted to match the PPM input signal rate. When a codeword is received, the IC logic checks it for timing and double-checks for possible code errors (by comparing consecutive frames) before translating it into a particular control function.

Figure 9.16 shows the practical circuit of a simple IR receiver using one of the ML926 to ML929 range of ICs. Here, the PPM input signals are fed to pin 3 via a pre-amp circuit, and the oscillator timing is controlled via $C_1-R_1-RV_1$ (and should be adjusted via RV_1 so that its periodic time is 1/40th of the time of a 0 interval of the PPM signal). The decoded four-bit binary output signals can either be used directly to give four output channels, or can be further decoded via a 4515B CMOS IC to give up to sixteen output channels.

transmitter code EDCBA	function	four-bit binary output DCBA
0000X	programme 1	0000
0001X	programme 2	0001
0010X	programme 3	0010
0011X	programme 4	0011
0100X	programme 5	0100
0101X	programme 6	0101
0110X	programme 7	0110
0111X	programme 8	0111
1000X	programme 9	1000
1001X	programme 10	1001
10100	analogue 1 +	*Note*
10101	programme step +	X = don't care digit
10110	analogue 2 +	
10111	analogue 3 +	
11000	standby	
11001	toggle O/P	
11011	normalize	
11100	analogue 1 −	
11101	programme step −	
11110	analogue 2 −	
11111	analogue 3 −	

Figure 9.18   *Basic twenty-one-command set for the ML922 receiver*

Finally, *Figure 9.17* shows the circuit of a versatile IC receiver with three digital and three analogue outputs, plus a four-bit binary output. R_3–R_4–RV_1 control the oscillator timing, C_2–R_3 give power-up reset action, and C_1–R_1 control the timing of the PB_1 manual-stepping facility. The table in *Figure 9.18* shows how the transmitter PPM codes relate to the twenty-one functions and the four-bit output codes of the ML922 IC in the above circuit.

The Siemens IR60 system

System basics
The basic Siemens system is designed to give up to sixty-channels of non-simultaneous remote control via a six-bit code transmitted along an IR beam. The system comprises one sixty-channel transmitter IC (the SAB3210), one infra-red pre-amplifier IC (the TDA4050), and one general-purpose receiver/decoder IC (the SAB3209) that provides three analogue and three digital outputs plus a four-bit parallel output and a six-bit serial output. Additional ICs include the SAB3211, which can convert the four-bit parallel output into a channel number display via a nine-segment (seven plus two segments) LED display, and the SAB3271, which is a complete receiver/decoder IC but gives, among other things, a six-bit parallel output.

The SAB3210 transmitter IC
Figures 9.19 and *9.20* show the outline and block diagram of the SAB3210, an IC specifically designed to transmit a carrier-modulated pulse-coded IR

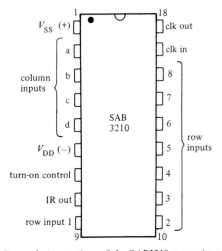

Figure 9.19   *Outline and pin notations of the SAB3210 transmitter IC*

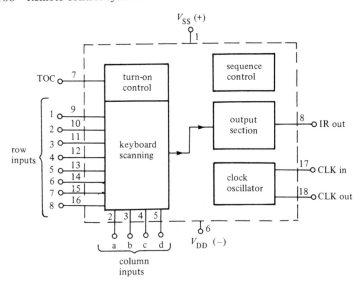

Figure 9.20   *SAB3210 transmitter IC block diagram*

signal. *Figure 9.21* shows a basic application circuit in which this IC receives input instructions via an eight-row (inputs 1 to 8) by four-column (inputs a to d) keyboard switch matrix: each key can thus be allocated a unique instruction number related to the matrix numbers (2_d, 8_a, etc.). Whenever a key is operated the IC detects the action and generates a start-bit plus a unique six-bit serial code word that is transmitted via the IR output stage; note that the IC's negative supply line is connected via external transistor Q_1, which is activated via turn-on control pin 7.

To input an instruction to the SAB3210 a column input must be connected to a row input, thereby turning the IC on via pin 7 and Q_1 and causing a corresponding carrier-modulated pulse code signal to be transmitted via pin 8 and Q_2 and causing a corresponding carrier-modulated pulse code signal to be transmitted via pin 8 and Q_2; carrier modulation occurs at half of the clock oscillator frequency (typically about 60 kHz) set by C_2–L_1–C_3. The pulse code signal continues to be transmitted as long as an input switch is closed. When the switch is released, this state is detected, and a unique end-command (code 111110) is transmitted.

When the IC is used in the simple manner shown in *Figure 9.21*, in which each code word corresponds to a one-row/one-column matrix combination, a total of thirty-two basic code words (000000 to 011111) can be generated. An additional twenty eight words (100000 to 111011) can be generated by using (with the aid of steering diodes) a two-row/one-column key combination in

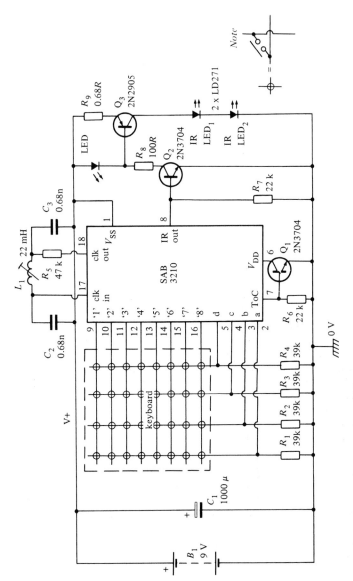

Figure 9.21   *Simple thirty-two-channel IR transmitter circuit*

which 8 always forms one of the two rows, as shown in the 60-key circuit of *Figure 9.22*. Here, two diodes are needed for every four additional instructions, with each diode pair connected to one other row and a set of column switches, so that actuation of a key connects two rows to one column; in practice, circuit reliability can be enhanced by externally biasing row 8 (and all other used rows) high via a 220 k resistor wired directly between the positive supply line and the relevant row pin.

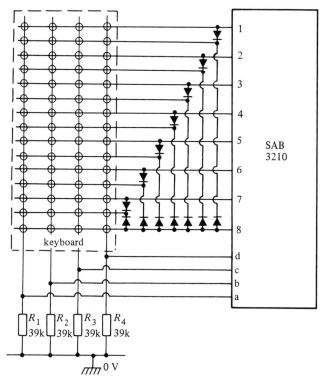

Figure 9.22   *Modification to give sixty-channel transmitter operation*

The SAB3210 transmitter IC is specifically intended to be used in conjunction with the SAB3209 receiver IC, and the table in *Figure 9.23* shows the relationship between the transmitter key codes and the receiver actions, using the basic thirty-two-channel instruction set. *Figure 9.24* shows the twenty-eight additional actions that can be obtained by using the 60-channel instruction set of *Figure 9.22*. Note that the SAB3210 IC can in fact generate a total of sixty-four different serial codes, but that four of these codes (instruction numbers 60 to 63) cannot normally be used.

inst no.	key code	serial code fed CBA	function	receiver output action
0	1a	000 000	normal position/on	volume sets to ⅓, analogue 1 and 2 set to ½. RLA on
1	1b	001	quicktone	volume sets rapidly to zero
2	1c	010	standby	turns RLA off
3	1d	011	reserve 1	RSV_1 output switches alternately high or low
4	2a	100	programme step +/on	increments binary channel-select output by one step : turns RLA on
5	2b	101	programme step −/on	decrement binary channel-select output by one step : turns RLA on
6	2c	110	on	turns RLA on
7	2d	111	reserve 2/on	RSV_2 output switches alternately high or low : turns RLA on
8	3a	001 000		
9	3b	001		
10	3c	010		
11	3d	011		not evaluated by SAB 3209 receiver, but read out at its serial interface
12	4a	100		
13	4b	101		
14	4c	110		
15	4d	111		
16	5a	010 000	channel 1 / on	sets binary output to 0000 : RLA on
17	5b	001	channel 2 / on	sets binary output to 0001 : RLA on
18	5c	010	channel 3 / on	sets binary output to 0010 : RLA on
19	5d	011	channel 4 / on	sets binary output to 0011 : RLA on
20	6a	100	channel 5 / on	sets binary output to 0100 : RLA on
21	6b	101	channel 6 / on	sets binary output to 0101 : RLA on
22	6c	110	channel 7 / on	sets binary output to 0110 : RLA on
23	6d	111	channel 8 / on	sets binary output to 0111 : RLA on
24	7a	011 000	channel 9 / on	sets binary output to 1000 : RLA on
25	7b	001	channel 10 / on	sets binary output to 1001 : RLA on
26	7c	010	channel 11 / on	sets binary output to 1010 : RLA on
27	7d	011	channel 12 / on	sets binary output to 1011 : RLA on
28	8a	100	channel 13 / on	sets binary output to 1100 : RLA on
29	8b	101	channel 14 / on	sets binary output to 1101 : RLA on
30	8c	110	channel 15 / on	sets binary output to 1110 : RLA on
31	8d	111	channel 16 / on	sets binary output to 1111 : RLA on

Figure 9.23 *Relationship between transmitter key codes and receiver actions on the thirty-two basic instructions*

Transmitter waveforms

The SAB3210 transmitter IC is a very sophisticated device and generates a fairly complex output signal, with the general form shown in *Figure 9.25*. Whenever a keyboard switch is operated the IC checks that only a single

inst no,	key code	serial code fed CBA	function	receiver output action
32	81a	100 000		
33	81b	001	–	
34	81c	010	–	
35	81d	011	–	not evaluated by SAB 3209 receiver, but read out at its serial interface
			–	
36	82a	100	–	
37	82b	101	–	
38	82c	110	–	
39	82d	111	–	
40	83a	101 000	volume +	increases volume output level
41	83b	001	volume –	decreases volume output level
42	83c	010	analogue 1 +	increases analogue 1 output level
43	83d	011	analogue 1 –	decreases analogue 1 output level
44	84a	100	analogue 2 +	increases analogue 2 output level
45	84b	101	analogue 2 –	decreases analogue 2 output level
46	84c	110	analogue 3 +	not evaluated by SAB 3209, but available on SAB 4209 receiver
47	84d	111	analogue 3 –	
48	85a	110 000	–	
49	85b	001	–	
50	85c	010	–	
51	85d	111	–	
52	86a	100	–	not evaluated by SAB 3209 receiver, but read out as its serial interface
53	86b	101	–	
54	86c	110	–	
55	86d	111	–	
56	87a	111 000	–	
57	87b	001	–	
58	87c	010	–	
59	87d	111	–	
60	–	100	–	not used
61	–	101	–	not used
62	–	110	–	'END' instruction
63	–	111	–	not permitted, due to ambiguity

Figure 9.24 *Relationship between transmitter key codes and receiver actions on the EXTENSION instructions*

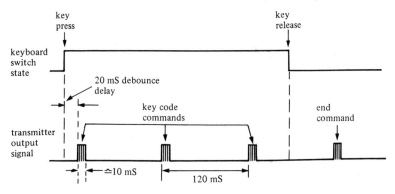

Figure 9.25   *General form of the SAB3210 transmitter signals*

switch has been closed and then, after a 20 mS debounce delay, transmits a suitably coded command signal. This signal has a typical duration of 10 mS and is repeated at 120 mS intervals so long as the keyboard switch is closed; when the switch is released a final end command signal (code 111110) is generated, and all transmission then ceases.

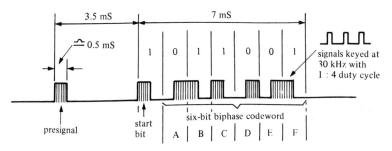

Figure 9.26   *Details of command signal with six-bit codeword 100110 and 1 start bit*

Figure 9.26 shows the general details that apply to each command signal (assuming the use of a 60 kHz clock oscillator frequency), but with specific details related to the six-bit codeword 100110 (reading from F to A). The command starts with a 0.5 mS pre-signal pulse which simply advises the receiver circuitry that a code signal is about to be transmitted. Roughly 3.5 mS later a 0.5 mS start bit pulse is transmitted, and is followed by six 0.5 mS pulses which form the six-bit codeword by using a so-called biphase modulation technique. What happens here is that the transmitted code signal carries imaginary markers at 1 mS intervals; a 0.5 mS pulse immediately following a marker represents a logic-1 bit, but a 0.5 mS pulse starting 0.5 mS after a marker represents a logic-0 bit. Thus, the biphase signal shown in *Figure 9.26* equals (reading from F to A) codeword 100110.

Note from *Figure 9.26* that each transmitted 0.5 mS pulse signal is modulated at 30 kHz (half of the clock oscillator frequency), with a 1/4 duty cycle, i.e., so that the pulse is switched high for one quarter and is switched low for three quarters of each 30 kHz cycle period. Thus, since each command signal comprises eight 0.5 mS pulses (giving a total command pulse period of 4 mS) and a 1/4 duty cycle is used, each command signal is switched high for a total of only 1 mS. Also, since each command is repeated at intervals of 120 mS, it can be seen that the *mean* current consumption of the IR transmitter output stage equals only 1/120th of the *peak* transmitter output current. This IR transmitter system is thus highly efficient.

Pre-amplifiers
In the IR receiver unit, the transmitted IR signal must first be detected and amplified via a suitable pre-amplifier circuit before being fed to the input of the SAB3209 receiver/decoder IC. This pre-amplifier circuit must be a fairly sophisticated design; it must be frequency-selective (to specifically respond to the IR code signal tone) and must provide very high signal gain (to give good long-range operation), but must not saturate when the transmitter is held very close to the receiver. *Figures 9.27* to *9.29* show the practical circuits of three suitable designs. Note that each of these pre-amplifiers must be mounted within a screened case, to minimize interference from unwanted electromagnetic signals.

The *Figure 9.27* design is based on a CA3140 op-amp and a number of readily available discrete components. Here, the basic IR signal is detected via

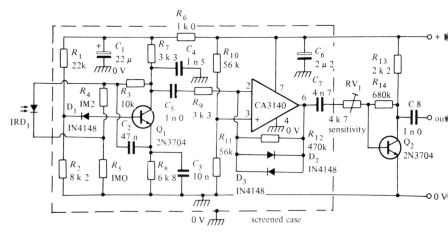

Figure 9.27   *IR receiver pre-amp circuit*

infra-red diode IRD_1 and is then selectively amplified via Q_1, the op-amp, and Q_2. To protect the circuit against signal saturation, $R_1-R_2-D_1$ and C_2 prevent the Q_1 bias point from shifting under heavy drive conditions, and D_2-D_3 clip the levels of the op-amp output signals, to prevent overdriving of following stages. The values of $C_2-C_3-C_4-C_5$ and C_7 are chosen to make the pre-amplifier reasonably frequency selective (to 30 kHz), thereby ensuring a good low-noise figure.

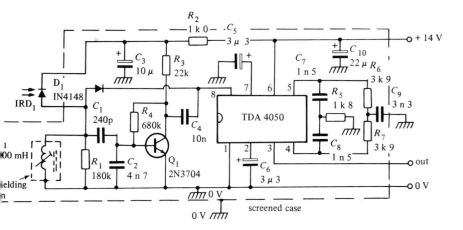

Figure 9.28   *TDA4050 IR receiver pre-amp circuit with* L-C *tuning*

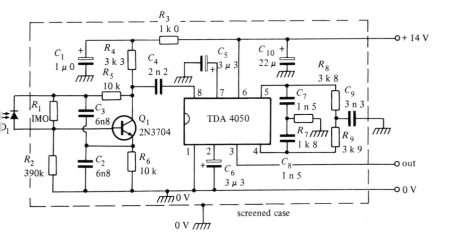

Figure 9.29   *TDA4050 IR receiver pre-amp circuit without* L-C *tuning*

The *Figure 9.28* and *9.29* designs are based on the TDA4050, an eight-pin dedicated IC that provides high overall signal gain (up to 100 dB) combined with an integral AGC system that is automatically activated (to eliminate saturation problems) by the pre-signal pulse in each transmitted command signal. In use, the TDA4050 is made frequency selective by wiring an external twin-T *R–C* network between pins 4 and 5, as shown. In the *Figure 9.28* pre-amplifier circuit the IR signals are first detected via IRD_1 and amplified via frequency-selective stage Q_1 (which is tuned via L_1–C_1–C_2) before being fed to the pin 8 input terminal of the TDA4050. In the *Figure 9.29* design the IR signal is detected via IRD_1 and fed non-selectively to the TDA4050 via the Q_1 stage.

The SAB3209
The SAB3209 acts as a general-purpose receiver/decoder IC in the IR60 remote control system. It provides three analogue and three digital outputs plus a four-bit parallel output and a six-bit serial output, all activated via the IR transmitter system. *Figure 9.30* shows the outline and pin notations of the IC (which is housed in an eighteen-pin DIL package), and *Figure 9.31* shows a practical application circuit of the device.

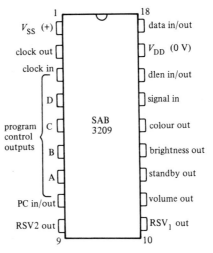

Figure 9.30   *Outline and pin designations of the SAB3209 receiver IC*

Input signals to the SAB3209 are fed to pin 15 via a suitable pre-amplifier stage. The IC is provided with a clock oscillator (L_1–C_1–R_6–C_2) which must be tuned to the transmitter clock frequency (double the serial code frequency). The chip checks the input code signal for sense (correct number of bits, bit duration, etc.), processes it and then both dumps the resulting code signal at

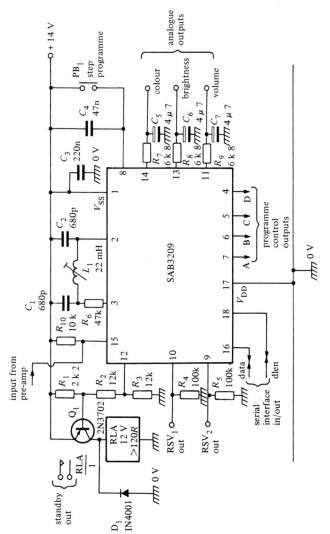

Figure 9.31   Practical SAB3209 application circuit

the serial interface (pins 16 and 18) and simultaneously passes it on to a register, from which it is then converted into a useful output action.

The outputs from pins 4 to 7 of the IC form a four-bit binary code that can be used to select (via suitable external decoder circuitry) any one of sixteen external channels; the binary code number can either be set via the transmitter signal, or can be directly shifted one step at a time via push-button switch PB_1 (connected to pin-8).

The outputs from pins 9, 10 and 12 are single-bit signals that can be set high or low via transmitter commands: in *Figure 9.31* the pin 12 output is used to activate relay RLA via Q_1, and the contacts of this relay can be used to switch power to external circuitry.

The SAB3209 provides three analogue output signals (at pins 11, 13 and 14); in *Figure 9.30* these outputs are notated *volume*, *brightness* and *colour* respectively, but in practice they can be used to control (via suitable external circuitry) any analogue functions. Each of these analogue outputs takes the form of a 1 kHz (approximately) square wave that can have its mark/space ratio (and thus its mean level) varied over a full span in sixty-four discrete steps via the transmitter command signals. These variable mark/space signals are converted to DC analogue voltages via low-pass filters $R_7–C_5$, $R_8–C_6$ and $R_9–C_7$, and can be used to control external voltage-controlled attentuators, amplifiers and filters, etc.

It should be noted that the SAB3209 receiver IC can be activated both via remote control signals fed into input pin 15 and by command signals fed directly into the serial interface (pins 16 and 18) terminals. If both types of signal are input simultaneously, those at the serial interface are automatically given priority by the chip logic.

Special support chips

Siemens produce a number of special (for professional use) support chips for use with the IR60 system. One of these is the SAB3211 display decoder/driver chip, which gives a visual readout of the selected binary-coded channel number on a multisegment LED display; this IC simply converts the four-bit output code from pins 4 to 7 of the SAB3209 into a form suitable for driving a nine-segment numeric display.

Another special IC is the SAB3271, a self-contained receiver/decoder chip which specifically decodes the transmitted six-bit serial command code signal into the form of six parallel outputs, which can be decoded by additional circuitry to make all sixty command instructions available for external use.

Finally, UK readers wishing to experiment with the IR60 remote control system should note that they can, in case of difficulty, obtain the Siemens range of ICs from Electrovalue Ltd, 28E St Jude's Road, Engelfield Green, Egham, Surrey TW20 0HB.

Index

179

Integrated circuits (IC) and components by type number